A Survey of Islamic Astronomical Tables

E. S. Kennedy

American Philosophical Society
Independence Square Philadelphia

This Second Edition of E. S. Kennedy's *A Survey of Islamic Astronomical Tables* is subsidized by a fund in memory of Margot Neugebauer. The first edition of this work appeared in *Transactions of the American Philosophical Society*, Vol. 46, Part 2 (1956).

Library of Congress Catalog Card No. 56-10045
International Standard Book No. 0-87169-462-X.

Preface

In the course of the thirty-three years that have elapsed since the publication of the survey, a great deal of work has been done in the field. In particular, Professor David King and the author have been assembling materials for the further study of these documents. The number of named zijes presently known stands at about 220, double the number listed in the survey. Of these, 122 are extant, well over half of those whose names are known. We have microfilms of twenty-nine, a quarter of those extant. Nine zijes have been published by various authors, a notable increase over the two noted in the survey. Detailed descriptions of six more have appeared, and descriptions of twenty-five others are in draft.

Present plans contemplate separate publication of the material under three categories:

(1) A list giving for each known zij, where possible, author, epoch date, an indication of its contents, significance, relations with others, and bibliography.

(2) A list of parameter sets giving, where possible, mean motions, initial positions, maximum planetary equations, apsidal initial longitudes, plus a sexagesimal index.

(3) Detailed descriptions of unpublished individual zijes, to be turned out in batches as occasion offers.

However, in spite of the accumulation of additional information, the original survey retains considerable utility. It describes the typical organization and contents of zijes in general. It defines the functions and operations, many of them peculiar to ancient and medieval astronomy and astrology, which are displayed as numerical tables in the zijes. This material continues to be valid, and may serve as an introduction to readers newly approaching the Islamic exact sciences.

The survey closes with a summary and analysis of the data assembled in it. In most respects the conclusions still hold. The general list is incomplete rather than erroneous. Hence it seems worthwhile to reprint the entire work at this time.

E. S. K.

A SURVEY OF ISLAMIC ASTRONOMICAL TABLES

E. S. Kennedy

CONTENTS

1. INTRODUCTION

The most impressive aspect of the source material for the study of medieval oriental astronomy is its overwhelming quantity. Thousands of Byzantine Greek, Sanscrit, Hebrew, Arabic, Persian, and Turkish astronomical and astrological manuscripts exist, many in uncatalogued collections, and most of them untouched by modern scholarship. In the case of the Arabic material a good deal of work has been done by Europeans, and the broad outlines of the subject have been well delineated. But anyone wishing to assist in building up a precise and detailed picture of Islamic astronomy is constrained to choose his material from amid a welter of easily available manuscripts.

Of these manuscript masses it is possible to isolate a fairly well-defined group of works, the *zījes*, which in the opinion of the author, make up the most significant and historically rewarding subclass of the whole. A *zīj* consists essentially of the numerical tables and accompanying explanation sufficient to enable the practising astronomer, or astrologer, to solve all the standard problems of his profession, i.e. to measure time and to compute planetary and stellar positions, appearance, and eclipses. These handbooks sometimes, but not always, contain explanations and proofs of the theory and reports of the observations on the basis of which the tables were computed. But in all cases the tables themselves, as the end results of theory and observation, can be used to reconstruct the underlying geometric models as well as the mathematical devices utilized to give numerical expression to the models.

This paper is a survey of the number, distribution, contents, and relations between zījes written in Arabic or Persian during the period from the eighth through the fifteenth centuries of the Christian era. Section 2 below discusses the etymology of the word and its introduction into various languages. Section 3 is a list of all such zījes extant or known to have been extant. Section 4 introduces a general classification of the subjects standard to these handbooks, together with such definitions and notations as will be found useful in the sequel. Each of the next twelve sections is a detailed abstract of a particular zīj, chosen for its importance or availability. The concluding sections, 17 and 18, point out developmental trends in these documents and indicate lines which future study may take. For ready reference an index of authors and zīj titles follows the bibliography.

2. ETYMOLOGY

It is well established that the word *zīj* (Arabic plurals *azyāj*, *zījāt*, and *ziyaja*), like a number of other technical terms, came over into Arabic from Persian. The stock explanation [1] is that the Persian cognate originally meant a thread or cord, in particular a bowstring, whence a chord in the geometric sense. In modern Arabic a mason's line is called a *zīj*. By extension the word came to stand for the set of parallel threads making up the warp of a fabric. Thence, from the resemblance between the closely drawn vertical lines of a numerical table and the warp set up in a loom the meaning was further extended to include the former. And by a final extension the word came to denote whole sets of astronomical tables, which is the meaning we use. Presumably this evolution had already taken place in Sasanian times, for there is a Pahlavi (Middle Persian) reference[2] to the *zīk-i shatroayār* (= *zīj-i shahriyār, cf.* §3, **30** below). In Islamic times, however, the word was sometimes still used to denote individual tables, as with **X200** and **X201** in §3 below.

In the Persian and Arabic dictionaries two Persian words are given as the source of *zīj*. Some have *zīg*,

[1] See, for example, *Nallino, Batt.*, vol. I, p. xxxi; *Khwar.*, p. 32; A dictionary of the technical terms used in the sciences of the Musalmans, Part II. Calcutta, 1862, p. 610; Murtaḍā al-Ḥuseinī al-Ḥanafī az-Zubeidī, "Tāj al-'arūs," Part II, Cairo, 1306 (A.H.) p. 55. Complete titles of references made in italics may be found by consultation of the bibliography at the end of the paper.

[2] West, E. W., Pahlavi texts translated, vol. IV (Sacred Books of the East, vol. 37), Epistles of Mānuščihar, II, ii, 9–11; Introd., p. xlvii.

from which the transition to *zīj* is natural. Others say, as does Bīrūnī in the Masudic Canon (Treatise III, Chapter 1, *cf. Schoy, Mas'ūdī*, p. 1), that it is from *zih*, which indeed means *cord* in the modern language. Perhaps it would be more correct to say that *zīg* is the Middle Persian form of a word from which New Persian *zih* developed.

For present purposes our basic criterion of a zīj shall be that the document in question contain a more or less complete set of tables of the topics listed in §4, regardless of whether demonstrations of the underlying theory are or are not included. This crosses the classification of Muslim astronomical works made in *Nallino* (p. 119), but it conforms to medieval usage. In some cases our choice may have been inconsistent. Thus we include the "Almagest" of Abū Naṣr Manṣūr (§3, **77**) under the assumption that it probably contained independent tables, but we exclude Naṣīr ad-Dīn aṭ-Ṭūsī's redaction of the Almagest on the ground that his independent Zīj-i Īlkhānī (§3, **6**) is extant.

The Greek word κανών,[3] in meaning very close to *zīj*, has likewise been Arabicized, as *qānūn*, and the two words are sometimes used interchangeably. Thus the *Handy Tables* of Theon are known either as *Zīj Thā'ūn* (§3, **X205**) or simply as *al-Qānūn*. In the same manner al-Bīrūnī called his zīj *al-Qānūn al-Mas'ūdī* (§3, **59**).

From Arabic or Persian the word *zīj* entered Byzantine Greek as ζήζι[4] and medieval Latin as *zich*, or *ezich*[5] from *az-zīj* the form with the definite article.

3. A GENERAL LIST

Since no date can be assigned to many of the works listed below, they are ordered and numbered arbitrarily. The bold-face serial numbers thus assigned are retained for identification throughout the remainder of the paper. Where approximate dates can be given they are noted, and in the case of extant works the location of one or more manuscript copies is given. No attempt has been made to cite all the available manuscript copies, especially in the case of the later, popular zījes, where the number of extant copies is occasionally large.

It is assumed that the original language is Arabic unless stated to the contrary.

While the general picture of Islamic astronomy which emerges from this study is accurate, it is inevitable that many errors in detail should eventually turn up in it. The situation as to authorship and dating is, in the case of many zījes, extremely complicated. The same zīj may be known by several different names, or documents thought to be independent may turn out to be the same. Sources frequently give conflicting information. In cases where copies of the works can still be found, many of these anomalies will eventually be resolvable by examination of the actual manuscripts; in many other instances the inconsistencies will remain. The list follows:

1. *Az-Zīj al-Mushtamil* of Aḥmad bin Muḥammad *an-Nihāwandī*, *c.* 790, who carried out observations at Jundīshāpūr, a scientific center in southwestern Iran from Sasanian times, non-extant. Ibn Yūnis (**14**) states that he knew of no observations of the solar mean motion in the interval between those of Ptolemy and the Mumtaḥan (**51**) save the observations of an-Nihāwandī.

(*Suter*, p. 10; *Caussin*, pp. 154–157.)

2. *Az-Zīj 'alā Sinī al-'Arab* (The Zīj (computed) According to the Arab (i.e. Hijra) Years), by Ibrāhīm bin Ḥabīb . . ., abū Isḥāq *al-Fazārī*, *c.* 750, of Baghdad, non-extant, was based on the Sindhind, **28** below.

There are numerous references to this work in the writings of *Bīrūnī*. In two places (*Risā'il*, I, pp. 133, 156) he speaks of it as al-Fazārī's Sindhind Zīj, once as *as-Sindhind al-Kabīr*. In the *India* (p. 208) he refers to the zīj of al-Fazārī and Ya'qūb ibn Ṭāriq (**71**) as though it were a single work. Perhaps this zīj is to be identified with **28** and **71** below.

See also **45** below.

A set of al-Fazārī's planetary parameters are tabulated in §17.

(*Nallino*, pp. 48 and 213–215; *Suter*, p. 1; *Fihrist*, p. 381.)

3. *Az-Zīj al-Waqibiya* (?) is an apochryphal title of the zīj of *Jamāl ad-Dīn* abī al-Qāsim bin Maḥfūẓ, al-Munajjim *al-Baghdādī*, *c.* 920 (?), extant as *Paris* 2486. According to *HKh*, the author chose what seemed to him the best elements of a number of zījes and incorporated them into his own.

(*HKh*, vol. III, p. 559; *Suter*, p. 197.)

4. The *Zīj-i Ashrafī* (in Persian) is three times cited by *Taqīzādeh* (pp. 162, 300, and 366) as having been written *c.* 1310 by Muḥammad bin abī 'Abdallāh *Sanjar al-Kamālī*, Seif-i Munajjim of Yazd (?) in central Iran. In Blochet, E., "Catalogue de la collection des Mss. orientaux . . . formée par M. Charles Schefer . . .," Bibl. Nat., Paris, 1900, p. 114, the manuscript given by Taqīzādeh as Supplement Persan 1488 is called *Tārīkh-i* (i.e. History of) *Ashrafī*. It is described as a treatise on astronomy, however, and is doubtless a zīj.

5. *Az-Zīj al-Amad 'alā al-Abad*, one of three zījes by Abū al-'Abbās Aḥmad bin Yūsuf, *ibn al-Kamād*, *c.* 1130, of Spain, the elements having been deduced from the *Toledan Observations* by az-Zarqālī. The work is non-extant as such, but see **24**, **48**, **66**, and **72** below.

(*HKh*, vol. III, pp. 568–569; *Suter*, p. 196; *Nachtr.*, p. 185; *GAL*, Suppl. vol. I, p. 864.)

[3] *Cf.* Liddell and Scott, A Greek-English lexicon (New Edition), Oxford, 1940; vol. I, p. 875.
[4] *CCAG*, vol. I, p. 3; vol. 5, p. 145; *Nallino, Batt.*, vol. I, p. xxxi.
[5] *Khwar.*, pp. 1, 32; *Wüstenfeld*, p. 21.

6. The *Zīj-i Īlkhānī* (in Persian) (abstracted in §13 below) by the famous *Naṣīr ad-Dīn aṭ-Ṭūsī*, *c.* 1270, in collaboration with a group of scientists gathered at the observatory at Marāgha in northwestern Iran under the patronage of the Īlkhan Hūlāgū, grandson of Genghis Khan. Other zījes mentioned in this work are **14**, **51**, **55**, and **70**. Numerous copies, commentaries, and an Arabic translation of this work are extant, that used for the abstract being *Bodl. Pers.* 1513 (Hunt 143). Part of this zīj, from the commentary of Maḥmūd Shāh Khuljī, was published in Latin by John Greaves (Latin, *Gravius*) as "Astronomica quaedam ex traditione Shah Cholgii Persae," and as "Binae Tabulae Geographicae, una Nassir Eddini Persae, altera Ulug Beigi Tatari," both in London, 1652.

For a description of the instruments used at Marāgha, see *Seemann*.

(*GAL*, vol. I, p. 511, Suppl. vol. I, p. 931; *Suter*, p. 149.)

7. *Az-Zīj al-Bāligh* of *Kūshyār bin Labbān al-Jīlī* (i.e. of Gīlān, a region of northern Iran), *c.* 1010. *Taqīzādeh* states (p. 226) that this work is distinct from **9** below, which see. In **35** below Shams al-Munajjim indicates the same thing.

(*Krause*, p. 519.)

8. *Az-Zīj al-Badī'* is one of five zījes (see **67**, **78**, **79**, and **90** below) non-extant, produced by the Banī *Amājūr* (or *Mājūr*) *c.* 910, working in Baghdād and Shīrāz, the father, Abū al-Qāsim, his son Abū al-Ḥasan 'Alī, and the freedman of the latter, by name Mufliḥ. Their observations are cited by Ibn Yūnis in **14** below (*Caussin*, p. 152 and on numerous other pages). They compared their observations with results obtained by computation from one of the zījes (the "Arabic" one) of Ḥabash (**16**). 'Alī was known personally to Ibn al-Adamī (**18**), who thought highly of him. 'Alī reports that he made numerous observations of the maximum lunar latitude, that he found it in general to exceed the standard Ptolemaic value of five degrees, but he obtained different maxima for different observations. This speaks well for the precision of his observational technique.

(*Suter*, p. 49; *GAL*, Suppl. vol. I, p. 397; *Fihrist*, p. 390, Transl., p. 35; *Delambre*, p. 139.)

9. *Az-Zīj al-Jāmi'* (abstracted in §10 below) ($\stackrel{?}{=}$ *Az-Zīj al-Jāmi' wal-Bāligh*) of *Kūshyār ibn Labbān al-Jīlī*, *c.* 1010, is extant in Berlin and Leiden. The copy here abstracted is *Leiden* Ms. 1054 (Cod. 523(1) Warn.). A table of contents of the *Berlin* copy, Ms. 5751, is given in the catalogue, and comparison of this with the Leiden manuscript suggests a reason for the disagreement between the sources as to whether Kūshyār composed two zījes or only one.

Both manuscripts are organized in four treatises, these being (1) directions, (2) tables, (3) explanations, and (4) proofs. The last two treatises are lacking in the Berlin copy. Of the portions which are present in both manuscripts, the number of chapters is about the same and the titles almost identical, but the order of presentation is widely different in the two versions, The table titles likewise are pretty much the same, but the sine table of the Leiden version, for example, is much more extensive than that of the Berlin manuscript.

It is possible that the two different names denote rearrangements such as those noted above of a single basic document. *Bīrūnī* (e.g. *Risā'il*, II, pp. 42, 52, and 62) refers in numerous places to Kūshyār's Jāmi' Zīj. Since for other works he usually names the author but not the zīj, this may be to indicate that he means that particular zīj of Kūshyār and not another one.

As remarked in **44** below, the elements of this work have been taken from al-Battānī (**55**), and it is improbable that new observational data have been incorporated into it.

At the end of the Berlin version are a number of isolated tables attributed to various individuals, among them Ḥabash (**15**), al-Battānī (**55**), Ibn al-A'lam (**70**), Yaḥyā (**51**), and (Abū Ma'shar) al-Balkhī (**63**).

(*Suter*, p. 83; *HKh*, vol. III, p. 570; *GAL*, vol. I, p. 222, Suppl. vol. I, p. 397; *Krause*, p. 519.)

10. The *Zīj-i Jāmi'-i Sa'īdī*, written in Persian, was the work of an individual who was contemporary with Jamshīd (**20**) and Ulugh Beg (**12**). Ṭabāṭabā'ī (writing in the Iranian periodical *Āmūzesh va Parvaresh*, 1319 [Hijrī Shamsī], No. 3, p. 4) quotes a passage from the introduction to these tables which indicates that among astronomers generally there was dissatisfaction with the Īlkhānī Zīj (**6**) because of errors made in reducing the observational data.

11. *Az-Zīj al-Jadīd* (abstracted in §14 below) of *'Alā' ad-Dīn ibn ash-Shāṭir*, *c.* 1350, sometime *muwaqqit* of the Umayyad Mosque of Damascus. The work was apparently popular, several abridged forms of it having been prepared, and it remains extant in numerous copies. The one used for the abstract is *Bodl. II, 2*, Ms. 278 (Seld. A. inf. 30), and in this copy at least, the tables have been computed on the basis of observations taken in Damascus, not in Cairo, as reported in some sources.

(*GAL*, vol. II, p. 126, Suppl. vol. II, p. 157; *HKh*, vol. III, p. 557; *Suter*, p. 168.)

12. The *Zīj-i Jadīd-i Sulṭānī* = *Zīj-i Ulugh Beg* = *Zīj-i Gurgānī* = *az-Zīj al-Kurkānī* = *Zīj-i Mīrzā Ulugh Bīk* = *Zīj-i Sa'īd-i Jadīd-i Gūrgānī* (in Persian, abstracted in §16 below) of the prince of Samarqand, *Ulugh Beg*, *c.* 1440, grandson of Tamerlane. This zīj, like a number of others (e.g. **6**, **8**, and **51**) is the product of the joint efforts of a group of astronomers. It may well have been the most widely used of all zījes and is still extant in hundreds of copies. For our abstract *Bodl. I, 1*, Ms. LXX (Pocock. 226) has been used. Arabic and Turkish versions of this zīj were made, and several later works were based on its tables. The first part was translated and published by L. Sédillot as "Prolégomènes des tables astr. d'Oloug-Beg, . . .," Paris, 1847 (*Sédillot* in the bibliography). The star table of this zīj has been published by *Knobel*. There is an earlier and very rare edition of the star table by Thomas Hyde, "Tabulae longitudinis et latitudinis stellarum fixarum ex observatione Ulugh Beghi . . .," Oxford, 1665. Also rare is John Greaves, "Epochae celebriores, . . . ex traditione Ulug Beigi," London, 1650. See also **6**.

A recent publication of great interest is *Kary-Niyazov*, which contains a description of the excavations made at the site of Ulugh Beg's observatory, under the auspices of the Academy of Sciences of the Uzbek S. S. R.

(*GAL*, vol. II, p. 212, Suppl. vol. II, p. 298.)

13. *Az-Zīj al-Jadīd ar-Riḍwānī*, revised by *Quṭb ad-Dīn* Maḥmūd bin Mas'ūd bin Muṣliḥ *ash-Shīrāzī*, *c.* 1290, the revision being extant as *Berlin* Fol. 3902, although it is not listed in the catalogue.

(*GAL*, Suppl. vol. II, p. 297.)

14. *Az-Zīj al-Kabīr al-Ḥākimī* of 'Alī ibn . . . Aḥmad *ibn Yūnis* of Cairo, *c.* 990, is a famous work which is extant only in fragments. The one at *Leiden*, Ms. 1057 (Cod. Or. 143), has a table of contents for the entire work which lists a total of eighty-one chapters. Of these the first twenty are found in the Leiden manuscript. *Paris* Ms. 2495 is a copy of the Leiden fragment. However, *Bodl. II, 2*, Ms. 298 (Hunt 331) picks up where the Leiden copy leaves off, beginning with Chapter 21 and continuing through 44. Except for minor variants the chapter titles are those given in the Leiden table of contents. The chapters are badly out of order, presumably owing to the folios having been bound in disorder. The two fragments are written in different hands, so they cannot be parts of the same manuscript. Thus something over half the total number of chapters is available. Since these parts alone comprise some 240-odd folios it is evident that the zīj was well named *al-Kabīr*, the great.

Additional isolated chapters are to be found in *Paris* Ms. 2496, and *Escorial* I, 919, 5.

Portions of this work have been published and translated. The introduction, with the table of contents and Chapters 4, 5, and 6 are to be found in *Caussin*. The parts of the zīj relative to sundial theory have been studied by Karl Schoy and the results published as "Gnomonik der Araber," of *Die Geschichte der Zeitmessung und der Uhren, Band I, Lieferung F*, Berlin, 1923. The same author has translated and commented upon Chapter 10 (in *Schoy, Trig.*), dealing with the computation of the sine table. A part of Chapter 11 is translated in *Schoy, Geogr.* Schoy translated additional chapters, publishing them in the 1920, 1921, and 1922 volumes of "Annalen der Hydrographie und maritimen Meteorologie." The latter have not been seen by the present author.

The entirety of the Leiden fragment plus eighteen additional chapters was translated by J. J. Sédillot early in the nineteenth century. This translation was never published and eventually disappeared. The effort was not wholly wasted, however, for pp. 76–156 of *Delambre* are given over to a penetrating analysis of Ibn Yūnis' work prepared with the aid of Sédillot's manuscript. In the judgment of Delambre, Ibn Yūnis was inferior to al-Battānī as an observer, but superior to him in computational technique.

The usual topics are covered in this zīj, and in a manner not radically different from corresponding sections of other tables. What is of almost unique interest is the author's wide knowledge of the work of his predecessors, his acute critical faculty, and an attitude toward observational errors and computational precision which is modern in tone and which would be completely foreign, say, to Greek astronomy. Of particular utility to the historian of science is his opening apologia in which he builds up a case for the presentation of his own zīj. He gives evidence of having had at hand zījes **1**, **8**, **15**, **18**, **28**, **46**, **51**, **63**, **70**, **91**, and **92** when he composed his own. From many of these works he quotes passages, reproduces a large number of planetary mean motions and other parameters, and reports the results of many observations.

(*Suter*, p. 77; *Delambre*, pp. 76–156; *GAL*, vol. I, p. 224, Suppl. vol. I, p. 401.)

15. *The Berlin Ḥabash Zīj* (abstracted in §7 below) is the work of an individual of great interest, Aḥmad bin 'Abdallāh, *Ḥabash al-Ḥāsib* (the Computer) al-Marwazī (i.e. of Marv), *c.* 850, about whose person and accomplishments the fog of history has closed thickly. Nor is it much dispelled by examination of this manuscript, *Berlin* 5750, and the other extant zīj written by him, **16** below.

He was one of the astronomers contemporary with the Abbasid Caliph al-Ma'mūn, but he does not seem to have been one of the small group of four or five who collaborated in the actual Mumtaḥan (**51**) observations. Ibn Yūnis (**14**) reports observations of his made in Baghdad in 829 and in 864. In another place Ibn Yūnis, who gives the impression of being a severe but balanced critic, says that Ḥabash's remarks concerning the latitudes of Venus and Mercury sound like those of one who does not understand what he is saying. To Ḥabash, Ibn al-Qifṭī ascribes three zījes:

1. A re-working of the Sindhind (**28** below) but including the "trepidation" of the fixed stars.
2. The *Mumtaḥan Zīj* (**51**), the best known of his works and based on his own observations.
3. The *Small* (**39**), or *Shāh* Zīj (**30**?).

The *Fihrist*, however, credits him with only two, the *Damascus Zīj*, and the *Ma'mūnic Zīj*, while Ibn Yūnis in many places refers to determinations arrived at by use of the *Arabic Zīj* of Ḥabash. As both *Caussin* and *Taqīzādeh* indicate, this need not imply that he composed works in languages other than Arabic. It may only distinguish this zīj (the "Arabic" one) from others based on Hindu or Persian methods. Or better, the usage may be owing to the tables having been based on the Arabic (i.e., Hijra) calendar rather than on the Yazdigird or Seleucid eras. And in fact both **15** and **16** display the planetary mean motions tabulated according to the Hijra calendar.

In another place, discussing various determinations of the obliquity of the ecliptic, Ibn Yūnis speaks of Ḥabash's Mumtaḥan zīj, which, he says is called *al-Qānūn* (The Canon).

Ḥabash is frequently mentioned in the writings of *Bīrūnī* (**59**), but never with an indication that he wrote more than one zīj. In the *Risā'il* (II, p. 81), for instance, he gives an example and proof of Ḥabash's method of computing planetary equations. In the *Chron.* (p. 177) he recommends his tables of lunar visibility.

The Berlin manuscript opens with the phrase: "(Thus) spake Aḥmad bin 'Abdallāh, known as Ḥabash al-Ḥāsib, the author of this zīj." But the rest of the introduction involves only lofty generalities and makes no statement concerning Ḥabash and his contemporaries, in strong contrast to the introduction to **16**. Subsequently, in the body of the zīj, mention is made of an-Nairīzī (**46** and **75**) who lived after the time of Ḥabash. And the epoch for mean motions is 511 A.H., whereas Ḥabash was surely dead by 300 A.H. So this manuscript is at best a late rescension of the work of Ḥabash.

These reservations as to authorship having been accepted, the manuscript remains of great interest because of its numerous individual tables of an unorthodox character.

(*Suter*, p. 12; *Nallino*, p. 275; *Taqīzādeh*, p. 212; *Fihrist*, p. 384, transl., p. 29; *Caussin*, pp. 58, 70, 159, 160, 172, etc.)

16. The *Istanbul Ḥabash Zīj* (abstracted in §8 below) is (Istanbul) *Yeni Jami* Ms. 784, 2°. For references and information about the author see **15** above.

The introduction to this work is clearly written by Ḥabash himself. After the standard pious injunction he commences with the customary opening gambit of the zīj-writer—his friends have prevailed upon him to do it. He reviews al-Ma'mūn's sponsorship of astronomical research, stating that he ordered an examination of the available literature and mentioning the Sindhind (**28**), the Arkand (**X214**) and the Shāh Zīj (**30**). Superior to all of these was the Almagest, he says. Ḥabash states that after the death of Yaḥyā (**51**) the latter was succeeded by Khālid bin 'Abd al-Malik al-Marwazī (i.e., al-Marvarūdī, **97**) as head of the caliph's commission. Khālid continued at Damascus the observations begun at Baghdād.

Ḥabash makes no mention of other zījes written by himself. The material leans heavily on Ptolemy; at the same time there are many sections which are distinctly non-Ptolemaic. The whole manuscript merits extensive study, especially the part devoted to eclipse theory.

(*Krause*, p. 446.)

17. *Mukhtaṣar az-Zīj* (≠ **85** ≠ **105**), a compendium of zījes using the methods of the Sindhind (**28**), by Aḥmad bin 'Abdallāh . . . abū al-Qāsim, *ibn aṣ-Ṣaffār*, *c.* 1100, of Cordova. This may be extant in Arabic written in Hebrew characters as *Paris, Bibl. Nat.* (Mss. Orientaux, Cat. des Mss. Hebreux et Samaritans . . ., 1866, p. 203) Ms. 1102.

(*Suter*, p. 86; *Suter Nachtr.*, p. 169; Steinschneider, *ZDMG*, vol. 47, p. 363; *GAL*, Suppl. vol. I, p. 401.)

18. *Az-Zīj al-Kabīr* $\stackrel{?}{=}$ *Naẓm al-'Iqd* are tables based on the method of the Sindhind (**28**) begun by Muḥammad bin al-Ḥusein bin Ḥamīd, *ibn al-Adamī* of Baghdād and completed *c.* 920 after his death by his student al-Qāsim bin Muḥammad al-Madā'inī. Non-extant. This work was available to Ibn Yūnis (**14**).

There is some confusion about the author's name, it being given also as Abū 'Alī al-Ḥusein ibn Muḥammad al-Adamī.

(*Suter*, p. 44; *Nallino*, pp. 203, 210; *Caussin*, p. 128.)

19. The zīj of Aḥmad bin Dā'ūd, *abū Ḥanīfa ad-Dīnawarī* is non-extant. In one place *HKh* states that Abū Ḥanīfa's observations were carried out at Isfahan in the year 849. This squares with the commonly given date of his death as 895. But it is irreconcilable with *HKh*'s additional statement that Abū Ḥanīfa's patron was the Buyid Sultan Rukn ad-Daula, *c.* 950.

(*HKh*, vol. III, pp. 470, 558; *GAL*, vol. I, p. 123, Suppl. vol. I, p. 187; *Suter*, p. 31.)

20. The *Zīj-i Khāqānī fī Takmīl az-Zīj al-Īlkhānī* (in Persian, abstracted in §15 below) of *Jamshīd Ghīāth ad-Dīn al-Kāshī* (or al-Kāshānī) *c.* 1420, the first director of Ulugh Beg's observatory at Samarqand. The zīj is dedicated to Ulugh Beg. Jamshīd died in 1429, before the completion of **12**. In his introduction the author pays his respects to Naṣīr ad-Dīn, the author of **6**, but expresses dissatisfaction with much of this zīj. His own object is to correct the mistakes of **6**. It is extant in two copies, *India Office* 430 (Ethé 2232), and *Aya Sofya* (Istanbul) 2692. The copy in *Meshed* (vol. III, p. 33, Ms. Math. 102, General No. 5329) is a fragment only, consisting of fifteen folios. The statement in the *India Office* catalogue (p. 1220) to the effect that this zīj is the first edition of **12** is incorrect. Preparatory to an eventual critical edition, the present author has made an English translation of this work.

Each of the six treatises of the Khāqānī Zīj is organized in three parts: (1) an extensive technical glossary of terms used in the treatise, (2) explanation of operations, and (3) proofs of the foregoing operations. The technical glossary alone makes the work highly valuable.

(*HKh*, vol. III, p. 563; *Suter*, p. 173; *GAL*, vol. II, p. 211, Suppl. vol. II, p. 295; *Krause*, p. 510; *Knobel*, p. 92.)

21. The zīj of Muḥammad Ibn Mūsā *al-Khwārizmī*, *c.* 840 (abstracted in §6 below), is one of the only two zījes out of the entire lot which has been published. In the original Arabic the work is not extant, but Adelard of Bath's Latin translation of the revision of Maslama al-Majrīṭī (fl. 1000) has been published by Björnbo and Suter (*Khwar.* in the bibliography). The unique interest which attaches to this publication is due to the fact that much of the material in it is non-Ptolemaic. An estimate of the origin of Khwārizmī's methods is made in §17 below.

Bīrūnī (in *Risā'il*, I, pp. 128, 168) notes the existence of a book by al-Farghānī, a younger contemporary of Khwārizmī, criticizing the latter's zīj, and Bīrūnī himself demonstrates (in *Risā'il*, I, p. 131) an error in Khwārizmī's planetary equation theory. It is curious to note that in spite of the simultaneous existence of tables based on more refined theories, this zīj was used in Spain three centuries after it had been written, and thence translated into Latin.

Bīrūnī also states that a work was composed by Abū al-Faḍl ibn Māsha'allāh comprising the zījes of Khwārizmī and Ḥabash (**15**) in shortened form.

Cf. **106** below.

(*Ibn al-Qifṭī*, p. 326; *Suter*, p. 10; *GAL*, vol. I, p. 215, Suppl. vol. I, p. 381.)

22. *Az-Zīj al-Malikshāhī* of *'Umar Khayyām*, *c.* 1090, of Nīshāpūr in northeastern Iran, non-extant, apparently dedicated to the Seljuk Sultan Malikshah.

(*HKh*, vol. III, p. 570; *Taqīzādeh*, p. 174.)

23. *Az-Zīj az-Zāhir* = *az-Zīj az-Zāmir* (?) of Farīd ad-Dīn abū al-Ḥasan 'Alī bin *'Abd al-Karīm ash-Shirwānī* (or ash-Sharwanī), known as *al-Fahhād* *c.* 1150, is one of six zījes (**53**, **58**, **62**, **64**, and **84**), all non-extant, attributed to this individual. That a number of this magnitude of essentially different handbooks should have been produced by a single astronomer is improbable, but the list of zīj names is given by three sources, *HKh* (vol. III, p. 567), al-Fārisī (**54**, *Lee*, p. 255), and Shams al-Munajjim (**35**, *Krause*, p. 519). In no two of the three lists are all the names identical, but all variants can be reconciled as alternative readings (or misreadings) of the Arabic characters, as with the zīj of this article.

By Shams al-Munajjim the author's name is given as 'Alī bin 'Abd [al-Karīm ?] *al-Bākū'ī*. But this also is reconcilable with the above, for both Shīrevān and Bākū are in northern Azarbaijan. This may fix the place of the author's origin, but it has no necessary bearing on where or for whom he worked, and the sources say nothing as to this.

Most of what is known about al-Fahhād comes via al-Fārisī, and the article on the latter's zīj (**54**) may be consulted in this connection.

Al-Fahhād was highly critical of the findings of most of his predecessors. His own observations carried out over a period of thirty years, bore out the planetary positions when computed according to the elements of Ibn al-A'lam (**70**). Exceptions were Venus, for which he observed a (deferent) apogee differing from that of the sun, and Mercury, for which he regards the Ptolemaic observations and theory as the best.

His observations corroborated the results of Yaḥyā ibn abī Manṣūr (**51**), al-Marwarūdī (**97**), al-Jawharī (**99**), and Ḥabash (**15**) in so far as the mean motions of the sun and moon are concerned.

(*Suter*, p. 318; *Taqīzādeh*, p. 366.)

24. A zīj was prepared by Ibrāhīm bin Yaḥyā an-Naqqāsh, *ibn az-Zarqālī* (or rather *az-Zargālla* as *Millás Vallicrosa 2*, p. 15 ff. has shown) al-Qurṭubī, known in Christendom as *Azarchiel*, of Toledo, *c.* 1100. These tables are not extant in Arabic. There are, however, many manuscripts of a Latin work entitled *Canones Azarchelis in Tabulas Toletanas* . . . discussed in *Steinschneider, Études*. The name itself is confusing, for in an astronomical context *canon* and *tabula* mean the same thing. *Curtze* (p. 348) gives another form of the title: *Canones Arzachelis sive regule super tabulas astronomie*. *HKh* (vol. III, p. 569) speaks of *Toledan Observations* (*Al-Arṣād aṭ-Ṭulaiṭuliya*), the results of which were worked over by az-Zarqālī. There seems to have been a group of Arab and Jewish scientists under the leadership of one Qāḍī Sa'īd (*Suter*, p. 106) who made astronomical observations in Toledo, *c.* 1060. It is probable that the Toledan group produced a zīj which was revised by az-Zarqālī.

Delambre (pp. 175–179) gives an abstract of the Latin translation, from which it is deducible that the tables contain an undigested mixture of Hindu and Ptolemaic methods. On the one hand it has a set of tables, of Hindu origin, for computing oblique ascensions in terms of right ascensions (*cf.* §6, C below). This includes, incidentally, a table of sines computed on the basis of a fundamental circle of radius 150. On the other hand, it gives tables of oblique ascensions as such, as does the Almagest. The set of maximum planetary equations quoted by Delambre is the set used also by al-Battānī (**55**). *Knobel* (*Chron.*, p. 10), by use of an argument involving the constant of precession, satisfied himself that the epoch of the star table in the Latin translation is contemporary with az-Zarqālī.

Recent work on the subject is to be found in Chapter II of the extremely valuable monograph *Millás Vallicrosa 2* listed in the bibliography. The author notes in numerous instances the close correspondence between al-Khwārazmī's zīj (**21**) and the Toledan tables. The table of sines reproduced on page **44** from the latter is identical with that appearing on page 142 of the Sengupta translation of Brahmagupta's Khaṇḍakhādyaka (see **X218**). The Toledan tables contain tables for computing the "trepidation of the equinoxes."

See also **X213** below.

(*Suter*, pp. 106, 109; *GAL*, vol. I, p. 473, Suppl. vol. I, p. 862.)

25. *Az-Zīj as-Sulṭānī* (in Persian) apparently not **12**, attributed either to *Quṭb ad-Dīn* Maḥmūd bin Mas'ūd bin Muṣliḥ *ash-Shīrāzī* (of **13** above) or Muḥammad 'Alī bin Mubārakshāh Shams ad-Dīn Mīrak *al-Bukhārī*, *c.* 1300, extant in *Tehran* II, 184. This work may possibly be the same as **32**.

(*GAL*, Suppl. vol. II, p. 297.)

26. The zīj of Abū al-Qāsim Aṣbagh ibn Muḥammad *ibn as-Samḥ*, of Granada, *c.* 1010, non-extant, is reported to have been based on the methods of the Sindhind (**28**). Examination of the *Libros del Saber* . . . (edited by Rico y Sinobas, Madrid, 1863–1867), vol. III, p. 241–271, fails to substantiate Suter's conjecture that this treatise is from the Ibn as-Samḥ zīj.

(*Suter*, p. 85; *Nachtr.*, p. 168.)

27. *Az-Zīj as-Sanjarī* = *Az-Zīj al-Mu'tabar as-Sanjarī as-Sulṭānī* = *Jāmi' at-Tawārīkh li's-Sanjarī* (abstracted in §12 below) by Abū Manṣūr abū al-Fatḥ 'Abd ar-Raḥmān *al-Khāzinī*, *c.* 1120, a Greek freedman of a judge of Marv in Turkestan. The copy abstracted is *Vatican* cod. Arab. 761. Another copy is listed in *Brit. Mus. List* as Or. 6669. Extracts from this magnificent work are in the library of the Sipahsālār Mosque of Tehran, in the manuscript numbered 681. A compendium of it is in Istanbul as Hamidiye Ms. 859. The work is dedicated to the Seljuk Sultan Sanjar ibn Malikshah.

(*GAL*, Suppl. vol. I, p. 902; *Nallino*, p. 227; *Nallino, Batt.*, vol. I, p. lxvii; *Vat. V.*, p. 72; *Krause*, p. 487.)

28. *Az-Zīj as-Sindhind* is an Arabic translation, *c.* 770, of one of the Sanskrit Siddhāntas, probably the Brahmasiddhānta of Brahmagupta (*Bīrūnī*, *Risā'il* III, p. 27) made at the Baghdād court of the caliph al-Manṣūr. This zīj exerted a deep influence on Islamic astronomy, especially in Spain, even after the Almagest came to be known and its superiority appreciated. Bearing witness to this statement are zijes **2**, **17**, **18**, **21**, **26**, **31**, **39**, **45**, **46**, **71**, and **90**, computed by the methods of the Sindhind or strongly affected by them. Except for **21**, none of these are extant, and the recovery of any of them might do much to clarify the vexed but fascinating problem of transmission and interaction of Babylonian, Hellenistic, Indian, and Sasanian astronomy.

In the *Risā'il* (e.g. II, pp. 24, 28) *Bīrūnī* refers several times to the Sindhind, in one place saying it is evident that the material in it was not obtained by actual observation.

(*Nallino*, pp. 48, 203, and 225; *Suter*, p. 4; *Steinschneider*, *ZDMG*, vol. 24, p. 353.)

29. *Az-Zīj ash-Shāmil* is an anonymous work extant in three copies (*Paris* Ms. Arabe 2528; *Brit. Mus. II*, *2*, Ms. 395, 3 (Add. 7492. Rich.) and Florence, Pal. 289[95]) concerning the authorship of which all manner of complicated conjectures have been made. In his introduction the author of the zīj states that he has adopted the mean motions worked out by Abū al-Wafā' al-Būzjānī (**73**) and his collaborators. These mean motions, he says, had been falsely represented by the author of az-Zīj al-'Alā'ī (**42** or **84**) as the results of his own observations. But, having found the same mean motions in a work of Abū al-Wafā', the author of the Shāmil zīj utilizes them there, having himself first checked them by observations of conjunctions. *Cf.* **56** below.

The first chapter discusses the Malikī calendar, whence it is clear that the work is later than, say, 1100. Concerning the institution of this era the author makes only the vague statement that it was begun by one of the eastern kings. It would be strange indeed to find an astronomer in Iran, or in any part of the Near East, for that matter, writing on chronology and not fully aware that it was Malikshah of the Seljuk dynasty who inaugurated this calendar. Perhaps this zīj was written in Spain or North Africa.

The functions tabulated are the standard ones for most zījes and a cursory examination fails to show any novelties of theory or presentation.

A microfilm of a Utrecht manuscript marked *Cod. Or. Rheno-traject 23* has been supplied to the present writer by Dr. P. Voorhoeve of the University of Leiden. This is a fragmentary copy of a zīj, the folios of which have been bound in great disorder. The front and back of the book are both missing. Although this work is not identical with the Shāmil zīj the basic parameters for the planetary mean motions in the two works are identical. Since we now know that **42**, **56**, **81**, and **84** all used Abū al-Wafā's mean motion parameters, the Utrecht fragment may be from any one of these. Since it likewise mentions the Malikī calendar, it cannot be part of the work of Abū al-Wafā' himself.

Cf. **56** below.

(*Suter*, p. 227; *Nachtr.*, pp. 166–7; *GAL*, Suppl. vol. I, p. 400.)

30. The *Zīj ash-Shāh* = *Zīj ash-Shahriyār* = *Zīj Shahriyārān ash-Shāh*, non-extant, was a set of tables translated into Arabic *c.* 790 (perhaps by one Abū al-Ḥasan 'Alī bin Zīād at-Tamīmī) from a Pahlavi original called *Zīk-i Shatro-ayār*. The date of composition of the latter is moot, Nallino arguing for *c.* 640, while Taqīzādeh points to a statement in **59** that the century-earlier Sasanian king Khosro Anūshīrvān convoked an assemblage of astronomers for the purpose of *correcting* the (hence already existent) Zīj ash-Shahriyār. Nallino's attempt to show that its contents were essentially Hindu is also disputed by Taqīzādeh.

The influence of this Pahlavi zīj greatly resembled that of the Sindhind (**28**), being very strong before the introduction of the Almagest into the Islamic world, and persisting for centuries thereafter, through the time of az-Zarqālī in Spain, for example.

The writings of Bīrūnī contain many illuminating references to this work. For instance, he reproduces (in the *Risā'il*, II, p. 222) a passage from a zīj whose author was unknown to him in which a computation purports to determine the terrestrial distance between Babylon, the location for which the Shāh Zīj was computed, and *al-Qubba* (The Cupola, i.e., Ujain) the base location of the Sindhind (**28**). The computation itself is of little value, for in it Babylon is alleged to be in the fourth climate. And this, as Bīrūnī takes evident pleasure in pointing out, is like transporting Baghdād to Nīshāpūr. Nevertheless the location of Babylon is not inherently improbable, for the Sasanian capital at Ktesiphon is in fact near Babylon.

In the same work (*Risā'il*, II, p. 148) he reports a method, given in the Shāh Zīj, for determining the time of day by the length of shadows. Following this is a method used in the same zīj for computing rising times.

He states (*Chron.*, p. 6) that this zīj uses midnight epoch in contrast to the general practice of using noon.

Most significant of all, he observes (*Risā'il*, III, pp. 24–30) that, at least with regard to the planetary equations, the parameters and methods used in the Shāh Zīj do not differ from those of the Sindhind (**28**) because they were taken from the Hindus to the Persians. And Abū Ma'shar, Khwārizmī, and Ya'qūb ibn Ṭāriq in their turn used the same system.

Various planetary parameters of this zīj are displayed in §17 below.

(*Nallino*, pp. 50, 226 ff.; *Taqīzādeh*, pp. 79, 268, and 320.)

31. *Az-Zīj al-Mukhtari'* was composed by *al-Ḥasan bin aṣ-Ṣabbāḥ*, *c.* 870 (?) (perhaps to be identified with *al-Ḥasan bin Miṣbāḥ*, not the "Old Man of the Mountain" of Isma'īlī fame). *Bīrūnī* (*Risā'il*, II, p. 39) names the zīj in connection with his discussion of gnomon lengths used by various observers. Al-Ḥasan is reported to have determined the mean motions according to the methods of the Sindhind (**28**), whereas planetary equations in his zīj were computed from the Ptolemaic model.

(*Suter*, p. 19; *Nallino*, p. 224.)

32. The *Zīj-i Shāhī*, (= Az-Zīj Ash-Shāhī ≠ **30**) (in Persian?) by *Ḥusām-i* (or Ḥusām ad-Dīn) *Sālār* is mentioned by Shams-i Munajjim (**35**); hence the work was written prior to 1320. The author may be either one of two individuals described by *Suter*, p. 195, and the work may possibly be the same as **25**.

Ahlwardt (*Berlin*, vol. X, p. 219) attributes a zīj with this name to Naṣīr ad-Dīn aṭ-Ṭūsī (**6**), on what basis is not stated. He probably refers to a manuscript (*Paris*, Fonds pers. 173), by 'Alī Shāh bin Muḥammad bin Qāsim *al-Bukhārī*, 'Alā' al-Munajjim, *c.* 1301.

See **40**, which is an abridgment of a zīj having this name.

(*Krause*, p. 519; *GAL*, Suppl. vol. I, p. 844; *Sédillot*, p. xcix; *Suter*, pp. 161 and 227.)

33. The *Zīj ash-Shastka* = *Zīj-i Shastgāh* of Ḥusein bin Mūsā *al-Hurmuzī* (or al-Hurmuzdī), *c.* 1180, of the Persian Gulf region (?), extant, but probably incomplete, as *Meshed* (Shrine Library), Vol. III, Ms. 333, 108, consisting of eight folios only.

(*GAL*, Suppl. vol. I, p. 866; *Sédillot*, p. xcvix.)

34. The zīj of Shams ad-Dīn Muḥammad bin Muḥammad *al-Ḥalabī*, *muwaqqit* of the Aya Sofya Mosque in Istanbul, hence dating after the Turkish capture of the city in 1453. Being mentioned by *HKh* it cannot be later than, say, 1650, but may be too late for inclusion in this list. These tables are based upon the observations of Ibn ash-Shāṭir (**11**). Non-extant.

(*HKh*, vol. III, p. 566.)

35. *Az-Zīj al-Muḥaqqaq as-Sulṭānī 'alā Uṣūl ar-Raṣad al-Īlkhānī* = *Zīj-i Shams-i Munajjim* (in Persian) by Muḥammad bin 'Alī Khwāja *Shams al-Munajjim al-Wābiknawī*, and dedicated to the Īlkhān Abū Sa'īd Bahādur Khān, *c.* 1320, is extant as *Aya Sofya* (Istanbul) 2694. As the title indicates, this work uses the elements of **6** above. The author mentions zījes **6**, **7**, **9**, **23**, **32**, **42**(?), **49**, **53**, **58**, **62**, **64**, **65**, **84**(?), and **108** in describing his motives for computing his own. *GAL* (Suppl. vol. II, p. 297), apparently erroneously, attributes this zīj to Muḥammad 'Alī bin Mubārakshāh Shams ad-Dīn Mīrak *al-Bukhārī* (Samspuchari). *Cf.* **32** above.

GAL also states that a Greek translation of these tables is in Florence as Ms. Laurentianus, Plut. 28, Cod. 17. In the *CCAG* (vol. I, pp. 85–89) excerpts from this manuscript are to be found. *Suter* (p. 219, note 80) regards the Greek work as a revision rather than a translation. In addition to some of the zījes referred to in the paragraph above, the excerpts mention also **6**, **27**, **32**, **44**, **55**, **56**, **84**, and **108**. The same zījes are mentioned also in an excerpt from a Vatican codex quoted in the *CCAG* in vol. V, part 3, p. 146.

(*HKh*, vol. III, p. 566; *Krause*, p. 519.)

36. The *Zīj ash-Shams wa al-Qamar* of Shihāb ad-Dīn abū al-'Abbās Aḥmad bin Rajab bin Taibugha *al-Majdī*, *c.* 1420, of Cairo is extant in *Cairo* (p. 295).

(*GAL*, vol. II, p. 128; *Suter*, p. 175.)

37. The *Kitāb as-Zīj fī 'Ilm al-Falak* of Sheikh Muḥammad bin abī al-Fatḥ *aṣ-Ṣūfī* al-Miṣrī (i.e., the Egyptian), *c.* 1480, is a simplified version of the zīj of Ulugh Beg (**12**), for the longitude of Cairo, extant in *Cairo* (p. 233), *Gotha* 1379, 1, and elsewhere.

(*HKh*, vol. III, p. 566; *Suter*, p. 185; *GAL*, Suppl. vol. II, p. 159; *Gotha*, vol. III, p. 39.)

38. A zīj (non-extant) by Muḥammad bin Aḥmad bin Yūsuf *as-Samarqandī*, who observed in Samarqand *c.* 870, is reported in **14**, where a number of observations of this astronomer are recorded.

(*Suter*, p. 28; *Caussin*, pp. 150, 152, 166, etc.)

39. *Az-Zīj aṣ-Ṣaghīr* (The Little Zīj) is the third of the three attributed to *Ḥabash* al-Ḥāsib *c.* 850. See **15** and **16**. Non-extant.

40. *Az-Zīj al-Mulakhkhaṣ 'alā ar-Raṣad al-'Alā'ī*, ($\stackrel{?}{=}$**56**) by *Athīr ad-Dīn* Mufaḍḍal bin 'Umar *al-Abharī*, *c.* 1240, is an abridgment of a Zīj-i Shāhī, (probably **32** rather than **30**). Extant as *Buhar* Ms. 347. *Cf.* also **42** and **84**. This work is not the same as **29**, although both have the same *incipit*.

(*HKh*, vol. III, p. 565, vol. IV, p. 567; *GAL*, vol. I, p. 608, Suppl. vol. I, p. 844; *Suter*, p. 145; *Būhār*, p. 381.)

41. *Tāj al-Azyāj wa Ghunyat al-Muḥtāj* (The Crown of Zījes . . .) is the work of Abū 'Abdallāh Muḥammad (or Yaḥyā bin Muḥammad) *ibn abī ash-Shukr al-Maghribi*, *c.* 1250, and is extant as *Escorial II*, Ms. 932 (Cas. 927). *Nallino* (*Batt.*, vol. II, p. xiv) has shown by an examination of the geographical tables of this work that it cannot be identical with **108**. It is probable, however, that a son of Yaḥyā is the author of **108**.

(*Suter*, p. 156; *GAL*, vol. I, p. 474, Suppl. vol. II, p. 869; Renaud, *Isis*, XVIII, p. 172.)

42. *Az-Zīj al-'Alā'ī*. Reports concerning a work bearing this name are so numerous and conflicting that it seems justifiable to describe **84** below as a separate zīj, and there may be more than two sets of tables involved, although none are extant.

HKh (vol. III, p. 567) states that it is in Persian and is the work of Niẓām al-A'raj (al-Ḥasan bin Muḥammad . . . an-Nīshāpūrī al-Qummī, d. *c.* 1330) for one 'Alā' ad-Daula. (The Buyid monarch having this name died *c.* 1050.)

In another place *HKh* (vol. III, p. 566) ascribes a zīj of this name to Mu'ayyad ad-Dīn al-'Urḍī (ad-Dimishqī, c. 1240) or to 'Alā ad-Dīn an-Nīshāpūrī, or toal-Bīrūnī (**59**).

The discussion in **29** adds nothing save that the composition of the 'Alā'ī work follows Abū al-Wafā' (**73**) d.*c.* 1000.

The Byzantine Greek work discussed in **35** also mentions an 'Alā'ī zīj.

Suter (*Nachtr.*, p. 167) makes a number of conjectures as to the authorship of this zīj.

The manuscript listed in *Vat. V.* as Borgiani arabi 91,1 may contain excerpts from this work.

(*Suter*, p. 161; *GAL*, Suppl. vol. I, p. 869).

43. The zīj of *'Umdat ad-Dīn* is mentioned only in *HKh*, vol. III, p. 566.

44. *Az-Zīj al-Fākhir* is reported by al-Fārisī (**54**) as having been written by 'Alī *an-Nasawī*, doubtless Ibn Aḥmad, abū al-Ḥasan, *c.* 1030, who worked under Buyid patronage in central Iran. The work as such is not extant, but following the colophon of the copy of **9** abstracted in §10 are several tables copied from the Fākhir zīj. They confirm the statement of al-Fārisī that the mean motions of this zīj, as well as those of **9**, **65**, and **49**, are taken from al-Battānī (**55**).

(*Lee*, p. 259; *Suter*, p. 96; *GAL*, Suppl. vol. I, p. 390.)

45. A zīj is reported (by *Bīrūnī*, *India*, p. 209) as having been written by one *Muḥammad bin Isḥāq bin Ustādh Bundādh as-Sarakhsī* in which an error of al-Fazārī (**2**) and Ya'qūb bin Ṭāriq (**71**) was rectified. In another place *Bīrūnī* (*Chron.*, transl. p. 29) says that the same individual worked out planetary cycles in the style of the Hindus but on the basis of improved observational data. In various places in the *Risā'il* he discusses Sarakhsī's planetary parameters. In one such place (III, p. 23) he speaks of him as being among the *Aṣḥāb as-Sindhind*, i.e., the advocates of its methods.

46. *Az-Zīj al-Kabīr* (The Big Zīj) li'n-Nairīzī was one of two zījes produced by Al-Faḍl bin Ḥātim *an-Nairīzī* (i.e., of Nairīz, a town near Shīrāz) abū al-'Abbās, *c.* 900, both non-extant. The other is **75** below, which see. The former, at least, was based on the Sindhind (**28**). A passage from one of these zījes is quoted by Ibn Yūnis (**14**). See **63** below.

(*Suter*, p. 45; *Caussin*, p. 118; *Fihrist*, p. 389, Transl., p. 35.)

47. *Az-Zīj al-Muṣṭalaḥ* of Muḥammad bin Muḥammad *al-Fāriqī* al-Muḥāsib is cited by *HKh* (vol. III, p. 568), but with no indication as to the date or place of writing. *Caussin* (p. 23) states that Ms. Arabe 1144 of the Bibliothèque Nationale is catalogued as a copy of the tables of Ibn Yūnis, but that in reality it has only a few tables from the latter and is a part of the Muṣṭalaḥ Zīj, which, he says, seems to have been composed in the fourteenth century. The catalogue of 1883–1895 (*Paris*) does not list an astronomical work under the old number 1144.

HKh in another place (vol. III, p. 470) says that the Ḥākimī (*cf.* **14**) observations were made in Egypt in the year 250 A.H. (864/5 A.D.) and that the results were used in the Muṣṭalaḥ Zīj.

48. *Az-Zīj al-Kāmil*, I = *Al-Kāmil fī at-Ta'ālīm.* This is another case where a number of references all mention a zīj or zījes having the same name. Here at least three distinct works, seem indicated, and the remaining two have been numbered **49** and **82**.

48 is a composite of **5**, **66**, and **72** by Abū al-Ḥasan (or Abū Muḥammad) 'Abd al-Ḥaqq al-Ghāfiqī, *Ibn al-Hā'im* al-Ishbīlī (i.e. of Seville), *c.* 1280 (?) who claims to have corrected the errors of Ibn al-Kamād. The work is extant as *Bodl. II*, *2*, Ms. 285 (Marsh 618). From the table of contents in the Bodleian catalogue it seems that in this work the theory of the trepidation of the equinoxes is adopted.

(*HKh*, vol. III, p. 568; *Suter Nachtr.*, p. 185; *GAL*, Suppl. vol. I, p. 864.)

49. *Zīj-i Kāmil*, II. This work is mentioned by Shams al-Munajjim (**35**) as having been written by Abū ar-Rashīd *ad-Dānishī*. It is also cited by al-Fārisī (**54**) as having been based on the mean motions of al-Battānī (**55**). Diacritical points are lacking from al-Fārisī's spelling of the author's name, but it can be read as ad-Dānishī.

(*Krause*, p. 519; *Lee*, p. 259.)

50. The *Zīj al-Ma'mūn* is mentioned by *HKh* (vol. III, p. 567) without other information than the *incipit*, which differs from that of **51** below.

51. *Az-Zīj al-Ma'mūnī li'l-Mumtaḥan* = *Tabula Probata* = *Zīj ash-Shamāsiya* = *Az-Zīj al-Mujarrab al-Ma'mūnī* by *Yaḥyā ibn abī Manṣūr*, *c.* 810, is the famous set of tables resulting from the observations of an astronomical commission appointed by the Caliph Ma'mūn and headed by Yaḥyā. The *Escorial II* manuscript Codex Arabe 927 abstracted in §5 below purports to be a copy of this work, the only one extant. But, as will be seen from the abstract, only the first few folios can represent part of the Mumtaḥan zīj in its original form. The remainder is material of great interest, but from earlier and later sources, mostly unacknowledged.

The name Shamāsiya associated with this work is that of the quarter of Baghdād in which the Mumtaḥan observatory was located. *Probata* is the medieval Latin translation of *mumtaḥan*. (*Cf. Millás Vallicrosa*, p. 76.)

Ibn Yūnis (**14**) states that the Mumtaḥan observations were concerned only with solar and lunar motions, and not with the motions of the planets proper.

(*Suter*, p. 8; *Fihrist*, p. 384; *Ibn al-Qifṭī*, pp. 357–59; *Caussin*, pp. 58, 172.)

52. *Az-Zīj al-Maḥmūdī* by Hibitallāh bin al-Ḥusein bin Aḥmad abū al-Qāsim, Badī' az-Zamān *al-Aṣṭurlābī*, *c.* 1120, of Baghdād, is not extant.

(*Suter*, p. 117.)

53. *Az-Zīj al-Muḥkam* is the second of the works of *al-Fahhād*, *c.* 1150, discussed in **23** above, which see. Non-extant.

54. *Az-Zīj al-Mumtaḥan al-Muẓaffarī* = *az-Zīj al-Mumtaḥan al-Khazā'inī* = *az-Zīj al-Mumtaḥan al-'Arabī* by Muḥammad ibn abū Bakr *al-Fārisī*, *c.* 1260 is extant as *Cambr.* 508 which also contains an abridged version of the same zīj. It is described by *Lee* and is of unusual interest for a number of reasons.

The work is dedicated to the author's patron, al-Malik al-Muẓaffar, Yūsuf bin 'Umar, King of Yemen, whence the first name of the zīj. It was written for the royal treasury (*al-khāzan*), whence the second. The tables are computed for the latitude and longitude of the Yemen. They are based on the observations of al-Fahhād (see **23**, etc.).

The introduction to the work, after the manner of Ibn Yūnis (**14**), makes critical mention of no less than twenty-eight other zījes, to wit, **1**, **8**, **9**, **14**, **15**, **23**, **38**, **44**, **46**, **49**, **51**, **53**, **55**, **58**, **62**, **63**, **64**, **65**, **70**, **84**, **91**, **93**, **94**, **95**, **96**, **97**, **98**, and **99**, all on the authority of al-Fahhād. Of these, **93** through **99** have been found mentioned nowhere else in the literature, and for a number of those cited elsewhere the information given by al-Fārisī has confirmed or established matters of authorship and families of zījes. These items have been incorporated into the articles of the works they concern directly.

Al-Fārisī has taken over intact the mean motions of the latest of the six zījes written by al-Fahhād, **84**.

He has a chapter and a table on what *Lee* (p. 263) takes to be a comet (*al-kaid*) "one of the stars having a tail, not situated in the firmament of the stars: but having its place in the etherial heaven, below that of the moon." For other references to this topic, see §4, *O* and §12, *E* below.

(*HKh*, vol. III, p. 567; *Suter*, pp. 139, 218; *Nachtr.*, p. 175; *Taqīzādeh*, p. 366; *Cambr.*, p. 93.)

55. *Az-Zīj aṣ-Ṣābi'* (abstracted in §9 below) of Abū 'Abdallāh Muḥammad bin Sinān bin Jābir al-Ḥarrānī *al-Battānī* (known as *Albategnius* in medieval Europe), *c.* 900, of Raqqa on the upper Euphrates, is the second of the two zījes which have been published. The monumental edition of this work by C. A. Nallino, "Al-Battānī sive Albatenii Opus Astronomicum" (abbreviated in the bibliography as *Nallino, Batt.*), 3 vols., Milan, 1899–1907, is the foundation of the modern study of Islamic astronomy. Unfortunately the translation of the text and the immensely valuable notes and commentary are in Latin.

As for the zīj itself, it is strongly Ptolemaic, with very little Hindu influence, and no change in the basic theory. Battānī was evidently an excellent observer, however, and his tables show considerable sharpening of parameters. Two Latin translations were made of this zīj, one having been printed in 1537. Thus Battānī's work was influential in the development of European astronomy.

Bīrūnī (*Risā'il*, III, p. 63), in connection with the computation of planetary sectors, mentions an abridgment of this zīj made by Abū al-'Abbās al-Ḥawālfa'sī (?).

Battānī's elements were taken over to form the basis of **9**, **44**, **49**, and **65**.

(*Suter*, p. 45; *Delambre*, pp. 10–62.)

56. *Az-Zīj al-Athīrī* (≟**40**) may be the work of *Athīr ad-Dīn* al-Mufaḍḍal bin 'Umar *al-Abharī*, *c.* 1240, of Mosul who is known to have computed a zīj. The conjecture is based entirely on the similarity between the names. An extract of this zīj is extant as *Vat. V.* Borgiani arabi 91,1. This work is referred to in the preface of **81**, the author of the latter stating that he has used it in writing **81**. He says further that the author of **56** in turn made use of the elements of Abū al-Wafā' (**73**). As stated under **29**, the anonymous author of **29** likewise utilized Abū al-Wafā'. From this the author of *Brit. Mus. II, 2* (p. 188) infers that **56** and **29** may be the same work, a conjecture which cannot presently be disproved, but which can by no means be regarded as established.

(*Suter*, p. 145; *GAL*, Suppl. vol. I, p. 844.)

57. *Az-Zīj al-Mukhtār*, by one *Abū al-'Uqūl* of Cairo, is extant as *Brit. Mus.* (*Suppl.*) 768. Apparently it is made up of selections from other zījes. Brockelmann (*GAL*, Suppl. vol. I, p. 864) gives its date as *c.* 1200, on what authority is not evident. The author has made use of the zīj of Ibn Yūnis (**14**).

58. *Az-Zīj al-Mustaufī* = *Az-Zīj al-Mustawī* is the third of the six zījes attributed to *al-Fahhād*, *c.* 1150, discussed in **23** above, which see. Non-extant.

59. *Al-Qānūn al-Mas'ūdī* = *The Masudic Canon* (abstracted in part in §11 below) is the work of the great scientist and scholar *Abū Reiḥān al-Bīrūnī*, *c.* 1030. This is extant in several copies, among which are *Berlin* Ms. 5667, *Brit. Mus. Suppl.* Ms. 756 (Or. 1990), Istanbul Carullah Ms. 1498, etc.

The entire work comprises eleven treatises, of which the third has been translated and commented on by *Schoy* (*Mas'ūdī*). Translations or publications of additional odd chapters also exist and are noted in §11 below. As has often been remarked the publication of the entirety of this extremely important document would be an achievement of first importance for the study of Islamic science. Such an edition was being prepared by Max Krause, who was killed during the course of World War II. The Oriental Publications Bureau of Osmania University, Lallaguda, Hyderabad-Deccan has undertaken to publish the Arabic text, and it is to be hoped that this will soon appear.[6]

From references in his writings, it is clear that Bīrūnī had access to zījes **2**, **15**, **19**, **21**, **28**, **30**, **31**, **45**, **46**, **55**, **63**, **70**, **71**, **73**, **82**, and **100**, many of which are no longer extant.

(*Krause*, pp. 479–80; *Suter*, pp. 98–100; *GAL*, vol. I, p. 475, Suppl. vol. I, pp. 870–75.)

60. An incomplete copy of a zīj in the library of the Iranian Parliament (*Tehran* Ms. 181) may be based on the observations of Ḥāmid bin al-Khiḍr, Abū Maḥmūd *al-Khujandī*, *c.* 990, who worked at Rayy (near Tehran) under the patronage of the Buyid prince Fakhr ad-Daula. This zīj is written in Persian, and the epoch of the tables of mean motions is 600 Yazdigird, some two centuries after the death of al-Khujandī.

(*Suter*, p. 74.)

61. A zīj (non-extant) is reported written by an astrologer named *Ḥārith*, *c.* 850.

(*Suter*, p. 19; *Fihrist*, p. 388, transl., p. 34.)

62. *Az-Zīj al Mu'tadil* = *Az-Zīj al-Mu'addal*(?) is the fourth of the six zījes attributed to *al-Fahhād*, *c.* 1150, discussed in **23** above, which see. Non-extant.

63. The *Zīj al-Hazārāt* (The Zīj of the Thousands) and *Zīj al-Qirānāt w'al-Iḥtirāqāt* are by Ja'far bin Muḥammad bin 'Umar *abū Ma'shar* (*Albumasar*) al-Balkhī, *c.* 850, the astrologer whose works and fame spread throughout medieval Europe. The tables are no longer extant, although there are many references to them in the literature. *Bīrūnī* (*Chron.*, pp. 29–31) states that they embodied the notion that at the time of the creation all seven planets were in conjunction (*qirān*), and that at the expiration of 360,000 years this great conjunction would recur, whereupon a new cycle of creation would commence. For this idea, taken over from the Sasanian Persians, Bīrūnī expresses great contempt, saying that any competent astronomer could work out a proper value for such a cycle on the basis of contemporary observations. This was in fact done, he says, by Abū al-Wafā', (**73**) as-Sarakhsī (**45**), and himself.

In the *India* (p. 157) he says that Abū Ma'shar took the longitude of Kangdezh (Kangdiz), a legendary castle in the extreme east, as the base point for his geographical locations.

In another work (*Risā'il*, II, p. 39) Bīrūnī remarks that "among the most marvelous things" is that in Abū Ma'shar's tables he takes the gnomon length as six and two thirds feet, whereas in his operations he

[6] Note added in proof: since the above was written, the first part of this publication has in fact appeared. It comprises the Arabic text of the first four treatises of this zīj and is called: Al-qānūn'l-mas'ūdī, Vol. I, Osmania Oriental Publications Bureau, Hyderabad-Dn., India, 1954.

computes as though it were six and a half feet. Moreover an-Nairīzī (**46**) and al-Hāshimī (**82**) incorporated the mistake into their own zījes.

As for Abū Ma'shar himself, Bīrūnī says with heavy-handed sarcasm that he is of those "who excite suspicion against astronomers and mathematicians by counting themselves among them."

In these tables the epoch of the Deluge, which is actually the Hindu era of the Kaliyuga (epoch 3102 B.C.), is employed, and *Bīrūnī* (*Chron.*, p. 136) gives the rule for expressing dates according to it.

In a third work (*Risā'il*, III, p. 21) he discusses Abū Ma'shar's method of determining the planetary equations.

Cf. **106** below.

(*HKh*, vol. III, p. 558; *Fihrist*, p. 386; *Suter*, p. 28; *Hāshimī*.)

64. *Az-Zīj al-Mughnī* is the fifth of the six zījes attributed to *al-Fahhād*, *c.* 1150, discussed in **23** above, which see. Non-extant.

65. The *Zīj-i Mufrad* (in Persian) by Abū Ja'far Muḥammad bin Aiyūb bin Ḥāsib *aṭ-Ṭabarī*, *c.* 1230 is extant as *Browne*, 0. 1. Al-Fārisī (**54**) states that this work, like **9**, **44**, and **49**, used the mean motions of al-Battānī (**55**).

(*HKh*, vol. III, p. 568; *GAL*, Suppl. vol. I, p. 859; Strothmann, *Der Islam*, vol. XXI, p. 298.)

66. *Az-Zīj al-Muqtabas* was produced by *Ibn al-Kamād*, *c.* 1130 by combining his other zījes, **5** and **72**. The same elements were again worked over, this time by Ibn *al-Hā'im*, to produce **48**, which see. The statements of *HKh* in this connection (vol. III, pp. 568–569) are complicated and confusing and may imply that a second zīj called al-Muqtabas was formed from **48**.

(*Suter*, p. 196, *Nachtr.*, p. 185; *GAL*, Suppl. vol. I, p. 864.)

67. The *Zīj al-Mamarrat* is another of the works of the Banī *Amājūr*, *c.* 910; non-extant. *Cf.* **8**.

68. A zīj is reported as having been produced by Muḥammad bin 'Abdallāh bin 'Umar, *Ibn al-Bāzyār*, *c.* 850, a student of Ḥabash (**15**).

(*Fihrist*, p. 385; *Suter*, p. 16; *Ibn al-Qifṭī*, p. 286.)

69. The zīj of al-Ḥasan bin Aḥmad bin Ya'qūb abū Muḥammad *al-Hamdānī*, Ibn al-Ḥā'ik, *c.* 930, who was born and died in Yemen, was widely used in the latter country. Non-extant.

(*HKh*, vol. III, p. 570; *Suter*, p. 53; *Ibn al-Qifṭī*, p. 163.)

70. *Az-Zīj al-'Aḍudī*, the non-extant work of 'Ali bin al-Ḥusein abū al-Qāsim al-'Alawī, *Ibn al-A'lam* ash-Sharīf al-Ḥuseinī, *c.* 960, was greatly admired and continued to be used until the thirteenth century. It is mentioned by Ibn Yūnis (**14**) who cites observations of the author. Bīrūnī also mentions it, calling attention to the difference between the planetary equations of the Almagest and those of Ibn al-A'lam. The latter worked under the patronage of the Buyid Sultan 'Aḍud ad-Daula, whence the name of his zīj.

(*Suter*, p. 62; *Caussin*, pp. 154, 170; *Biruni*, *Risā'il*, II, p. 23, and III, p. 30; *HKh*, vol. III, p. 470.)

71. *Az-Zīj al-Maḥlul min as-Sindhind li-Darajat Daraja* by *Ya'qūb ibn Ṭāriq*, *c.* 770, is, as its name indicates, based on **28**, and with the table entries computed for one degree intervals of the arguments. Ya'qūb is reported to have been in attendance at the court of the Caliph al-Manṣūr upon the arrival there of Kankah(?) an Indian astronomer who supplied the material for the zīj. See **2** above. Sachau (*Bīrūnī*, *India*, transl., vol. II, pp. 311–313) is of the opinion that Ya'qūb's work *Kitāb fī Tarkīb al-Aflāk* (Compositio Sphaerarum) is to be identified with the zīj.

(*Suter*, p. 4; *Nallino*, pp. 215 and 221; Steinschneider, *ZDMG*, vol. 24, p. 332; *Hāshimī*; *Bīrūnī*, *India*, transl., vol. II, pp. 67–68; *Ibn al-Qifṭī*, p. 378.)

72. *Az-Zīj al-Kaur 'alā ad-Daur* is the third of the three zījes ascribed to *Ibn al-Kamād*, *c.* 1130. Non-extant as such, it is reputedly incorporated in **48**. See also **5** and **25**.

73. *Az-Zīj al-Wāḍiḥ* ($\overset{?}{=}$ *The Almagest of Abū al-Wafā'*) is the product of the observations of Muḥammad bin . . . al-'Abbās, *Abū al-Wafā' al-Būzjānī*, *c.* 970, and his collaborators working at Baghdād under Buyid patronage. The parameters thus deduced were sufficiently well thought of to be taken over by the authors of **29**, **42**, **56**, and **81**.

It has been suggested that Abū al-Wafā's zīj and his Almagest are to be identified, or that the former is the set of tables to go with the latter. *Bīrūnī* (in *Risā'il*), however, writes of the two as though they were separate works. In any event the question is academic, since the zīj is non-extant and the Almagest is available only in part. It is *Paris*, Ms. Arabe 2497, and includes in its 107 folios only the first seven treatises of the work. It has no tables, but contains material of interest for the history of trigonometry.

(*Suter*, p. 71; *Fihrist*, p. 394; Carra de Vaux, L'almageste d'Abū'lwéfa Albūzdjāni, *Jour. Asiatique*, 19 (1822), pp. 408–471.)

74. *Az-Zīj al-Mufannun* is mentioned only by *HKh* (vol. III, p. 570) with no indication of date or author.

75. *Az-Zīj aṣ-Ṣaghīr* (The Little Zīj) *li'n-Nairīzī* is the second of the two ascribed to *an-Nairīzī*, *c.* 900. Non-extant. See **46** for references. Ibn Yūnis states (*Caussin*, p. 74) that Nairīzī adopted the solar mean motion of Yaḥyā (**51**) as determined at Baghdād.

76. A zīj is reported by a single source written by Yūsuf bin 'Umar *al-Juhanī*, *c.* 1020, of Toledo. These tables may be the *Tabulae Jahen* translated by Gerard of Cremona (apparently not extant).

(*Suter*, pp. 96 and 214; *Wüstenfeld*, p. 66.)

77. *Al-Majisṭī ash-Shāhī*, i.e. the Almagest (dedicated to) the Shāh (Abū al-'Abbās 'Alī bin Ma'mūn of Khwārazm) was the principal work of *Abū Naṣr Manṣūr* bin 'Alī bin 'Irāq, *c.* 1000, of Khwārazm. The author of this important work was the teacher, friend, and scientific correspondent of al-Bīrūnī (**59**). The treatise itself is non-extant, except for a short extract, *India Office*, 734,2°.

(Krause, M., Die Sphärik von Menelaos aus Alexandrien in der Verbesserung von Abū Naṣr Manṣūr b. 'Alī b. 'Irāq . . . , *Abhand. der Gesellschaft der Wissenschaften zu Göttingen*, Philol.-Hist. Klasse, Dritte Folge, Nr. 17, 1936.)

78. *Az-Zīj al-Khāliṣ* is another of the five zījes produced by the Banī *Amājūr*. *Cf.* **8** and **67**.

79. *Az-Zīj al-Muzannar* is likewise one of the productions of the Banī *Amājūr*, *cf.* **8**, **67**, and **78**.

80. A certain *al-Khāqānī* (not to be confused with **20**), *c.* 1010, is reported by *Ibn al-Qifṭī* (p. 181) to have been the author of astronomical tables. No other information is available regarding him or the work.

(*Suter*, p. 95; Steinschneider, *ZDMG*, 24, p. 350.)

81. *Az-Zīj al-Jadīd al-Musammā Durr al-Muntakhib* is by an individual who calls himself *al-Qiss Qīriāqus* (the priest Cyriacus), a Christian convert to Islam. The work is extant as *Bodl. II, 2*, Ms. 274. It can be dated as *c.* 1480 since not only is the equivalent Yazdigird year 850 taken as epoch for the chronological tables, but the author gives an example of conversion between calendars in which he chooses this year.

The introduction to this zīj is strongly reminiscent of that of **29**, indeed a short passage occurs identically in the two works. However, this is not a copy of **29**. But Cyriacus says, as does the anonymous author of **29**, that he has utilized the work of Abū al-Wafā' (**73**), but through the intermediary of the Athīrī Zīj (**56**), after checking by independent observations. *Cf.* under **56**.

All of the standard topics and types of tables are present, and since several tables are computed for a latitude of $\varphi = 37;\ 30°$ the work will have been for the vicinity either of Tabrīz (in northwest Iran) or Samarqand (Turkestan).

82. *Az-Zīj al-Kāmil* (*li'l-Hāshimī*) by Muḥammad bin *'Abd al-'Azīz al-Hāshimī*, *c.* 950, is cited by *Bīrūnī* (*Chron.*, p. 315), in connection with his listing of the feasts of the Sabians. Non-extant, this zīj is not to be confused with **48** or **49**. See **63** above.

(*Suter*, p. 79; *GAL*, Suppl. vol. I, p. 386.)

83. *Az-Zīj al-Ikhtiyārī* is a work extant as *Rāmpūr*, I, Ms. 428,41. This, *GAL* (Suppl. vol. I, p. 844) ascribes to *Athīr ad-Dīn* (**56**), tentatively equating the zīj to **56**. Neither the zīj nor the Rāmpūr catalogue are available to the present writer, and the grounds for this ascription cannot be assessed by him.

84. *Az-Zīj al-'Alā'ī ar-Raṣadī* is the sixth of the zījes ascribed to *al-Fahhād*, *c.* 1150. Non-extant. See in particular **23**, also **53**, **58**, **62**, and **64**. The elements of this zīj were used by al-Fārisī (**54**), it being the last of the series and based on observations of al-Fahhād himself.

85. *'Umar bin Muḥammad bin Khālid* bin 'Abd al-Malik (al-Marwarūdī), *c.* 880 (?), himself an observer and a son and grandson of astronomers, is reported to have written a zīj compendium (*mukhtaṣar*) based on the methods of the Mumtaḥan observers. The author's grandfather was a member of al-Ma'mūn's astronomical commission and to him **97** is ascribed.

(*Suter*, p. 38; *Fihrist*, p. 386; *Ibn Al-Qifṭī*, p. 242.)

86. *Az-Zīj az-Zāhī* is an extract from az-Zīj ash-Shāh (**30**) made by Yaḥyā bin Muḥammad bin 'Abdān bin 'Abd al-Wāḥid, Abū Zakariya, *Ibn al-Lubūdī*, *c.* 1250, of Syria and Egypt. Non-extant. See **87** below.

(*Suter*, p. 146.)

87. *Az-Zīj al-Muqarrab* is a second zīj produced by *Ibn al-Lubūdī* (see **86** above), this one based on the Mumtaḥan (**51**) observations.

88. The *Zīj at-Tashīlāt* (The Zīj of Simplifications) by *Jamshīd al-Kāshī* (**20**) *c.* 1420, non-extant, is a work which Kāshī himself lists among his productions, in the introduction to his Miftāḥ al-Ḥisāb (*cf.* Luckey, P., "Die Rechenkunst bei Gamšīd . . . al-Kāšī," Wiesbaden, 1951, p. 6). As will be seen in the abstract of Kāshī's other zīj, ff. 142r through 156r of the India Office copy consist of tables for simplifying the determination of true planetary positions. In addition there is considerable explanatory text. It is reasonable to conjecture that the author, having worked out the methods these tables embody, before the writing of **20**, should have issued them as a self-contained work.

89. A zīj was prepared by Muḥammad 'Alī bin Shu'aib, Fakhr ad-Dīn abū Shujā', *Ibn ad-Dahhān*, *c.* 1170, who worked in Damascus in the service of the famous Saladin. Non-extant.

(*Suter*, p. 126.)

90. A version of the Sindhind (**28**) was prepared by the Banī *Amājūr*, *c.* 910, non-extant. For references see **8**.

91. A zīj (non-extant) is ascribed to the three *Banī Mūsā*, *c.* 850, sons of Mūsā bin Shākir, by Ibn Yūnis (**14**), who quotes numerous parameters and observations from their work. Al-Fārisī (**54**) also attributes a zīj to these brothers. *Cf.* **92** below.

(*Suter*, p. 8; *Caussin*, pp. 148–151; *Lee*, p. 252.)

92. A separate zīj (non-extant) by *Aḥmad bin Mūsā bin Shākir*, Abū al-Qāsim, one of the three brothers of **91** above, is also attested to by Ibn Yūnis (**14**) who quotes parameters found by Aḥmad individually. For references see **91** above.

93. A zīj (non-extant) is attributed to the famous *Thābit bin Qurra*, *c.* 870, by al-Fārisī (**54**). Thābit advocated the theory of the equinox trepidation. See **94** below. His table of visibility limits is reproduced by al-Khāzinī (**27**, *cf.* §12, L below). His table of right ascensions appears in **15** (§7, C below). Ibn Yūnis (**14**), quoting a letter of his, says his elements (i.e., basic parameters) are too well-known to need writing down by him. Later he cites five equinox observations of Thābit in Baghdād. Thābit's version of Ptolemy's "Planetary Hypothesis" has been published in German translation in *Ptolemaeus Opera*, II, Leipzig, 1907, pp. 71–145. For the Greek text of Book I, see pp. 70–106.

(*Lee*, p. 252; *Suter*, p. 34; *Caussin*, pp. 114, 146; *GAL*, vol. I, p. 217, Suppl. vol. I, p. 384.)

94. A zīj (non-extant) is attributed to *Isḥāq ibn Ḥunein* (not to be confused with Ḥunein ibn Isḥāq, his father), *c.* 880, by al-Fārisī (**54**). Isḥāq, best known as a physician and translator, made a translation of the Almagest which was revised by Thābit (**93**). It is possible that this is the "zīj" referred to. It is also possible that both individuals recomputed the tables to conform to their own or other post-Ptolemaic observations.

(*Lee*, p. 252; *Suter*, p. 39.)

95. A zīj (non-extant) is attributed to the celebrated maker of astronomical instruments, Ḥāmid bin 'Alī *al-Wāsiṭī*, *c.* 870, by al-Fārisī (**54**).

(*Lee*, p. 252; *Suter*, p. 40; *GAL*, Suppl. vol. I, p. 398.)

96. *Sanad ibn 'Alī*, *c.* 830, a Jewish convert to Islam, played a leading role in the Ma'mūn observations. He was the author of a zīj, no longer extant, which was still being used in the thirteenth century.

(*Ibn al-Qifṭī*, p. 206; *Suter*, p. 13; *HKh*, vol. III, p. 466; *Caussin*, pp. 56, 66, 67, 94.)

97. A zīj (non-extant) was written by Khālid bin 'Abd al-Malik *al-Marvarūdī* (or *al-Marwazī*, of Marv), *c.* 830, according to al-Fārisī (**54**) and *HKh*. The former adds that the values used by Marvarūdī for the lunar and solar mean motion were the same as those employed in **15**, **51**, and **99**, which is not surprising since the authors of all of these zījes collaborated in the Ma'mūn observations.

(*Lee*, p. 252; *Suter*, p. 11; *HKh*, vol. III, p. 466.)

98. A zīj, non-extant, is credited to Muḥammad bin 'Īsā, abū 'Abdallāh *al-Māhānī*, *c.* 860, by al-Fārisī (**54**). Ibn Yūnis (**14**) reports many of al-Māhānī's observations at Baghdād.

(*Lee*, p. 252; *Suter*, p. 26; *Caussin*, pp. 102–113.)

99. Al-'Abbās bin Sa'īd *al-Jawharī*, *c.* 830, participated in the Ma'mūn observations both at Damascus and Baghdād. By Ibn al-Qifṭī, *HKh*, and al-Fārisī (**54**) he is alleged to have written a zīj, now non-extant. Al-Fārisī states further that his determination of the solar and lunar mean motions was confirmed by the observations of al-Fahhād (**23**, see above).

(*Lee*, p. 252; *Suter*, p. 12;, *HKh*, vol. III, p. 466.)

100. A short excerpt from the zīj of *Abū 'Āṣim 'Iṣām*, *c.* 760, a freedman of the powerful Khālid bin Barmak is given by *Bīrūnī* (*Risā'il*, II, p. 93). The passage relates to a method for determining the meridian solar shadow length in terms of the length at equinox, and Bīrūnī shows that it is closely related to the methods used in Hindu zījes.

101. *Az-Zīj al-Hārūnī* is referred to by *Bīrūnī* (*Risā'il*, II, p. 159) who describes an operation contained in it in which Hindu type sine functions are used.

102. A zīj is reported written by *Hārūn bin 'Alī bin Yaḥyā* bin Abī Manṣūr (i.e. the grandson of the author of **51**) *c.* 860. These tables may be identical with **101** above.

(*Suter*, p. 34.)

103. *Az-Zīj al-Kāfī* was written by *'Uṭārid bin Muḥammad*, according to *Bīrūnī* (*Risā'il*, III, p. 85), who mentions the work in connection with a discussion of planetary theory.

(*Suter*, p. 67; *Fihrist*, transl., p. 33.)

104. In several places *Bīrūnī* refers (*Risā'il*, II, pp. 87, 108; III, p. 89) to the zīj of Abu Bakr *Muḥammad bin 'Umar bin al-Farrukhān*, *c.* 780, whose family was from the Caspian coastal region of Iran. In one place Bīrūnī says that Ibn 'Umar's work is intermediary between that of Abū Ma'shar and the Persians.

(*Suter*, p. 17.)

105. Regarding *Az-Zīj al-Mukhtaṣar* ($\neq$ **17** $\neq$ **85**) of *Abū Muḥammad as-Sanafī* (?), *Bīrūnī* (*Risā'il*, III, p. 23) states that the author's claim to have determined certain planetary parameters by observation is false.

106. A *Zīj Ja'far* has been remarked by *Wüstenfeld* (pp. 21–22) of which at least one fragment of a Latin translation exists. It is listed in the *Catalogue général des mss. des bibl. publiques de France, Paris, Bibl. Mazarine*, Paris, 1890, t. III, pp. 151–152, ms. 3642 (1258), fol. 82–90, with the title "Liber Ezich Iafaris el Kauresmy, per Adelardum Bathoniensem ex arabico in latinum sumptus." It has been assumed that the transliteration of al-Khwārizmī in the title implies that the author of **21** is intended, but Wüstenfeld points out that Ja'far is not among the names of Muḥammad ibn Mūsā. It is the name of Abū Ma'shar, the author of **63**, and Wüstenfeld suggests that the Latin translator may have confused Khwārizm with Khurāsān, Abū Ma'shar's place of origin. This is possible, but Ja'far is a very common name, and we list the work separately pending examination of the manuscript.

(*Suter*, p. 11.)

107. From the zīj of *'Abd ar-Raḥmān* bin 'Umar, abū al-Ḥusein *aṣ-Ṣūfī*, *c.* 950, Ibn Yūnis (**14**) cites the basic parameter for the solar mean motion. No other available source mentions this zīj, the author being best known for his star catalogue.

(*Suter*, p. 62; *Caussin*, pp. 154–155.)

108. The zīj of *Muḥī* al-Milla w'*ad-Dīn* Yaḥyā bin Muḥammad, bin Abī ash-Shukr *al-Maghribī* al-Andalūsī, *c.* 1280, is extant as *Meshed* (Shrine Library) Ms. 332(103); the author assisted Naṣīr ad-Dīn in the preparation of **6**. See **41** above.

(*Suter*, p. 155; *GAL*, vol. I, p. 474, Suppl. vol. II, p. 869.)

109. A zīj written by a certain *Ibn al-Masīḥ*, Abū al-Qāsim Aḥmad *al-Gharnāṭī* (i.e. of Granada) *c.* 1060, is reported by *d'Herbelot* as being "un fort gros volume." The name of the author has been found nowhere else in the literature.

SUPPLEMENTARY LIST

For the sake of completeness, and to avoid confusion with regard to nomenclature, the following titles are listed and indexed, but with an X preceding the serial number. They consist either of (*a*) sets of tables called zījes which are not really zījes in our sense, or (*b*) proper zījes but dating before the eighth century or after the fifteenth.

X200. *Zīj aṣ-Ṣafā'iḥ*, Tables (for the Laying out of Astrolabe) Plates, by Abū Ja'far *al-Khāzin*, *c.* 950.

(*GAL*, Suppl. vol. I, p. 387.)

X201. The name *Zīj aṭ-Ṭailasān*, has been used for tables involving the determination of the length of daylight. It is also used by *Bīrūnī* (*Chron.*, p. 132) in speaking of a chronological table. The word *ṭailasān* is a Persian one meaning *hood* or *cowl* and probably denotes tables formed of right triangular arrays of numbers (hence resembling triangular cowls) joined along the diagonals. This is the form of the Bīrūnī table.

(*Suter*, *Nachtr.*, p. 165.)

X202. *Az-Zīj li-'Arḍ Makka* was computed *c.* 1680.

(*GAL*, Suppl. vol. II, p. 487.)

X203. The *Zīj-i Muḥammad Shāh-i Hindī* = *Zīj-i Jadīd-i Muḥammad Shāhī* = *Zīj-i Rajah Jai Singh Sawā'ī* (in Persian) was finished *c.* 1730.

(*Bankipore*, vol. XI, p. 69.)

X204. *Zīj-i Shāhjihānī* was computed *c.* 1610.

(*Knobel*, p. 92.)

X205. *Zīj Thā'ūn* = *al-Qānūn* = *Qānūn al-Masīr* are common appellations for the Theon *Handy Tables*.

(*Hāshimī*, *Fihrist*, transl., pp. 21, 53; *HKh*, vol. III, p. 470.)

X206. The *Zīj al-Harqan* is mentioned by *Bīrūnī* (*India*, transl., vol. I, p. 228). From the little al-Bīrūnī says of it, *Nallino* (p. 226) concludes that the date of its composition was 742 and conjectures that *harqan* is an Arabic transliteration of the Sanscrit *ahargaṇa*, a word meaning the number of days between a certain epoch and a given time. (*Cf.* Schmidt, O., "On the Computation of the Ahargaṇa," *Centaurus*, 1952, p. 140.) This fits in with the statement of Sachau (*Bīrūnī*, *India*, transl., vol. I, p. xxxv; vol. II, p. 378) to the effect that this zīj was probably a handbook for conversion of Hindu, Arabic, and Persian dates.

In another work *Bīrūnī* (*Risā'il*, III, p. 26) quotes from the Harqan three (Arabic) mnemonic verses used in Āryabhata's method for the determination of the solar and lunar equations.

X207. By *Zīj Baṭlamyūs* is probably intended the *Handy Tables* of Ptolemy, as distinguished from the Almagest (*al-Majisṭī*).

X208. *Zīj-i Qasīnī* = The Tables of Cassini, in all probability the "Tables astronomiques . . ." of Jacques Cassini, Paris 1740.

X209. *Az-Zīj al-Mufīd 'alā Uṣūl ar-Raṣad al-Jadīd* of Riḍwān ar-Razzāz, *c.* 1710.

(*Princeton*, Garrett 1004, p. 316.)

X210. *Az-Zīj al-A'shārī ash-Shāhanshāhiya* (The Imperial Decimal Zīj) is a late Turkish work.

(*GAL*, Suppl. vol. III, p. 1290.)

X211. *Az-Zīj al-Hindisī* (The Geometric Zīj) of Abū Faḍl al-Ḥayyānī (?), *c.* 950, non-extant, we take to be other than an astronomical handbook.

(*Suter*, p. 67; *Fihrist*, p. 391.)

X212. *Az-Zīj al-Muthannā ash-Sharjī* was compiled *c.* 1660.

(*GAL*, Suppl. vol. II, p. 567.)

X213. *Al-Qānūn li-Ūmāniyūs* (The Canon of Ammonius?) is the name of a work extant (but incomplete in Arabic, *Munich* Ms. 853) in the redaction of az-Zarqālī (**24**).

Steinschneider (*Études*, p. 2) states that several

copies of a Latin translation of this canon are extant, e.g. Bodleian, Cod. Laud. 644,19. According to the Latin version, these tables were originally prepared for the daughter of Ptolemy, for the meridian of Antioch, and were based on the Egyptian calendar.

In the Munich catalogue the name of the author is given (in Arabic characters) as Ūmātiyūs, the original form of which is conjectured to be Eumathius. But this name occurs in no known astronomical context, and the deletion of a single dot from the letter *ta* converts it into a *nūn*, and the word into the form given in the title, a natural transliteration of Ammonius. Moreover, it is known that the fifth century Ammonius of Alexandria computed a canon (*CCAG*, vol. II, p. 182), and the mention of a daughter may be a garbled reference to Hypatia, the daughter of Theon.

Chapters III and IV of *Millás Vallicrosa 2*, consist of a critical edition of the text of the Munich manuscript plus a Spanish translation. The tables are transcribed from the Munich manuscript and from an old Spanish version (Ms. Arsenal 8.222), "Cánones de Humeniz."

This work is an almanac rather than a zīj; nevertheless its numerous tables would repay detailed analysis. Like az-Zarqālī's zīj (**24**) it includes both the standard tables of oblique ascensions and also (p. 225) a table of "Differences of Ascensions . . ." (*cf.* §6, C below).

(*Nallino, Batt.*, vol. I, p. xxxv, note 5.)

X214. The *Zīj al-Arkand* is a Sanscrit work which was early translated into Arabic, at or before the time of Ya'qūb ibn Ṭāriq (**71**), and was widely used; *Bīrūnī* (*India*, p. 206, transl., vol. II, p. 7; *Risā'il* II, p. 133, etc.) says it is from the *Kandakātik Zīj* (the *Khaṇḍakhādyaka*, *cf.* **X218**) written by *Brahamkūbta* (*Brahmagupta*). In *Risā'il*, II, p. 150 he gives a method of computing the length of daylight as a function of the season and the equinoctial shadow length of a gnomon, the method being from the Arkand.

Sachau (*Bīrūnī, India*, transl., vol. II, p. 339) objects to the identification of *arkand* with *Khaṇḍakhādyaka* and proposes *Āryakhaṇḍa* instead.

X215. The *Zīj an-Naṣrānī* (= *Zīj Kassawṭūh*) is a translation of the tables of Abraham ben Šemu'ēl Zacuto, *c.* 1490, the original being in Hebrew.

(*Vat. V.*, Ms. Vaticani arabi 963.)

X216. *Az-Zīj li'n-Nairain* (*The Zīj for the Two Luminaries*) is cited by *Bīrūnī* (*Risā'il*, I, pp. 126, 165) as having been written by *Abū Dā'ūd, Suleimān bin 'Iṣma*. Bīrūnī gives a numerical example of the method used in it for computing the solar equation. Since apparently it deals only with solar and lunar motions it is not included in the main list.

X217. The *Karaṇatilaka Zīj*, a Sanscrit work, is referred to by *Bīrūnī* (*India*, p. 300, transl., vol. II, pp. 206, 209, etc.). In the same work (transl., vol. I, p. 156) the author is given as Vijayanandin. In *Bīrūnī, Risā'il* (II, p. 136) he appears, in Arabic transliteration, as Bijayānand.

X218. *Az-Zīj Kandakātik al-'Arabī* (*The Arabic Khaṇḍakhādyaka*) is a work translated by *Bīrūnī* (*India*, p. 300, transl., vol. II, p. 208) for a native of Kashmir named Syāvabala (?). Bīrūnī complained of the poor quality of the previous translation, **X214**. An English translation of and commentary on the original Sanscrit has been published by P. C. Sengupta (University of Calcutta, 1934).

X219. The *Karaṇasara Zīj* is another Sanscrit work referred to by *Bīrūnī* (e.g. *India*, p. 241, transl., vol. II, p. 79). He gives the author (transl., vol. I, p. 156) as Vitteśvara, the son of Bhadatta (or Mihdatta?). In Arabic transliteration (*Risā'il*, II, p. 136) it appears as Wittisfara.

X220. The *Zīj-i Jāmi'* = *Zīj-i Intikhābī* ($\neq$**9**) is apparently a compendium *c.* 1460 of the work of Maḥmūdshāh Khuljī, who wrote a commentary on **6**. Extant as *Bodl. Pers.* Ms. 1522 (Greaves 6).

It is appropriate to mention additionally a class of works consisting of books not themselves zījes, but which are about zījes. Some five of these are listed below.

1. *Ghurrāt az-Zījāt* (Choice Parts of the Zījes) perhaps written by Abū Muḥammad an-Nā'ib al-Āmulī, was used by *Bīrūnī* (*India*, transl., vol. I, pp. xxxvi, vol. II, pp. 90, 388; *Chron.*, p. 13).
2. *Al-Fikr al-Wahīj fī Ḥall Mushkilāt az-Zīj* (Ardent Thoughts on the Solution of Difficulties with the Zīj), by Muḥammad abū Bakr al-Fārisī, the author of **54**. (*Cf. Lee*, p. 254; *Suter*, p. 139.)
3. *'Ilal az-Zījāt* (Difficulties of the Zījes) by 'Alī bin Sulaimān al-Hāshimī. A copy of this is *Bodl. II, 1* 879,4 (Seld. A. 11).
4. *El libro de los fundamentos de las Tablas astronomicas*, by Abraham ibn 'Ezra, listed in the bibliography under *Millás Vallicrosa*.
5. A Commentary on the zīj of Khwārizmī (**21**) was written by Aḥmad bin Muthannā ibn 'Abd al-Karīm. (See *Millás Vallicrosa*, pp. 51–52; Steinschneider, *ZDMG*, vol. 24, p. 353.)

Of these, the present writer has examined only 3 and 4. The former is the work of an individual whose technical background was very meager, whose innocence of mathematics is betrayed whenever he attempts to use it, and whose style is diffuse and rambling. Nevertheless it has been possible to glean from it several parameters which were used to confirm the same values obtained from other sources, and the narrative material helps to round out the general picture of pre-Islamic and early Islamic astronomy. In this latter sense 3 is probably typical of the lot.

4. CLASSIFICATION OF SUBJECT MATTER

Having listed all reported zījes, we now set about an analysis in considerable detail of twelve of these complicated documents which in some cases run to over two hundred folios. Preparatory to this it is thought well to set up a standard outline of the topics common to most zījes. To make easier the comparison of corresponding components in different sets of tables, the order of this master outline will be adhered to in all cases (except in §11), although it should be stated that in the actual manuscripts the order of subject matter is standardized only partially or not at all. The outline will also be used for defining functions encountered in most zījes, and for the adoption of standard notations.

In abstracting a zīj main emphasis has been laid on the examination of the numerical tables, largely because of the pressure of time. While the theoretical and explanatory sections have by no means been neglected completely, it is to be expected that a great amount of new matter would be extracted by detailed study of these passages.

As is the case with Babylonian and Greek astronomical works, the great bulk of the numerical material of all the zījes is displayed in sexagesimals, in a place-value system with base sixty and a zero symbol. In transcribing sexagesimals from the Arabic alphabetical numerals we will make use of the notation

$$32, 15; 3, 0, 57,$$

for example, to express the number

$$32\times60^1+15\times60^0+3\times60^{-1}+0\times60^{-2}+57\times60^{-3}.$$

Readers interested in an exposition of one of the powerful and elegant sexagesimal algorisms developed by the Islamic mathematicians may consult Luckey, P., "Der Lehrbrief über den Kreisumfang . . .," Berlin, 1953.

We will also make use of a mixed system, widely applied in the zījes, in which the integer part of a number is represented decimally, the fractional part sexagesimally. Sexagesimal representation of the integer parts of numbers greater than sixty is rare, but sometimes observed.

The superscript *s* will be used to denote zodiacal signs. The subscript *s*, however, will stand for *sun*, *m* for *moon*, *n* for *ascending node*. The superscripts *h* and *d* are for hours and days respectively. The word *places* in contexts indicating the precision of a computation will always indicate *sexagesimal places*. In the description of individual numerical tables it is to be assumed that columns of tabular differences are not present unless mention is made to the contrary.

It has been found natural to adopt fifteen main divisions of subject matter, each indicated by a capital letter. Appropriate subdivisions are used as needed. The outline follows:

A. CHRONOLOGY

All zījes begin with one or more chapters and sets of tables devoted to the definition of the various eras and calendars in use at the time and place of writing, to methods of converting dates from one calendar into another, and to the problem of determining the *madkhal* or initial week-day of a given year and month in a given calendar.

The three most common calendars, and which are discussed in most zījes, are the Muslim (Hijra), the Seleucid, and the Yazdigerd (Persian). The latter, using the 365 day Egyptian year without intercalary days, is particularly convenient. Less commonly treated eras and calendars are the Jewish, Coptic (Era of Diocletian), Maliki (Seljuk), Soghdian, Hindu, and, in regions subject to Mongol rule or influence, the Chinese-Uighur. For descriptions of most of these calendars see *Ginzel*, vol. I.

Lists of holidays for various religions are frequently given as well as royal canons of ruling and extinct dynasties.

The Hindus and the pre-Islamic Arabs made use of twenty-eight prominent stars or groups of stars situated near the ecliptic for keeping rough track of the lunar month. These *lunar mansions* (*manāzil al-qamar*) are tabulated in some zījes. For a full discussion of the topic see *Nallino*, pp. 175–194.

B. TRIGONOMETRIC FUNCTIONS

All zījes contain tables of pure mathematical functions having no essentially astronomical character.

The most common of these are:

The Sine (*al-jaib*) which had replaced the Ptolemaic chord function by the time of composition of the earliest Muslim zījes. The function tabulated in the zījes (abbreviated as *Sin* to distinguish it from the ordinary *sin*) differs from the modern sine function only to the extent that the fundamental circle used in the definition has a radius of sixty (1, 0) instead of unity. The identity relating the two functions is therefore

$$\text{Sin}\,\theta \equiv 1{,}0 \sin\theta$$

for all θ. If a certain sin θ is expressed in sexagesimals the corresponding Sin θ is found by moving the sexagesimal point (;) of the former one place to the right. Thus

$$\sin 1° = 0;1,2,49,43,11,14,44,16,19,16,\cdots,$$

and

$$\text{Sin}\,1° = 1;2,49,43,11,14,44,16,19,16,\cdots$$

correct to ten places. For an exposition of the ingenious and powerful algorism by which this extraordinarily precise determination was obtained see Aaboe, A., "Al-Kāshī's Iteration Method for the Determination of Sin 1°," *Scripta Mathematica*, vol. 20 (1954), pp. 24–29.

The *Tangent* (*aẓ-ẓill*, literally *the shadow*, or *aẓ-ẓill al-awwal*, or *aẓ-ẓill al-ma'kūs*, or *aẓ-ẓill al-mankūs*) is defined analogously

$$\text{Tan}\ \theta \equiv 1{,}0 \tan \theta.$$

The same names were also used for the function

$$k \tan \theta$$

where, depending on the table, k was variously taken as $6\frac{1}{2}$, 7, or 12.

Analogous definitions and tables of the *cotangent* (*aẓ-ẓill ath-thānī*, the second shadow; *aẓ-ẓill al-mabsūṭ*, *the* (horizontally) *extended shadow*, etc.) were used. That these functions had as origin the lengths of shadows cast on a horizontal or vertical plane by a gnomon of length k is clear from the nomenclature.

The *versed sine* (abbreviated *Vers*; *sahm*, *sagitta*, i.e., *arrow*) defined as

$$\text{Vers}\ \theta \equiv 1{,}0 - \text{Cos}\ \theta \equiv 1{,}0 - \text{Sin}\ (90° - \theta)$$

was occasionally tabulated.

The *secant function* (*quṭr aẓ-ẓill*, the *shadow's hypotenuse*) was rarely tabulated.

C. SPHERICAL ASTRONOMICAL FUNCTIONS

For the solution of problems in spherical astronomy, transformations of coordinates, time-measurement, and so on, it is useful to have, in addition to tabulated trigonometric functions, tables of functions which are

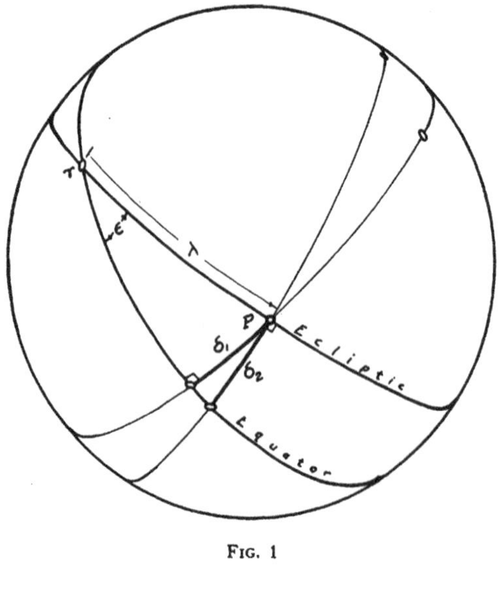

FIG. 1

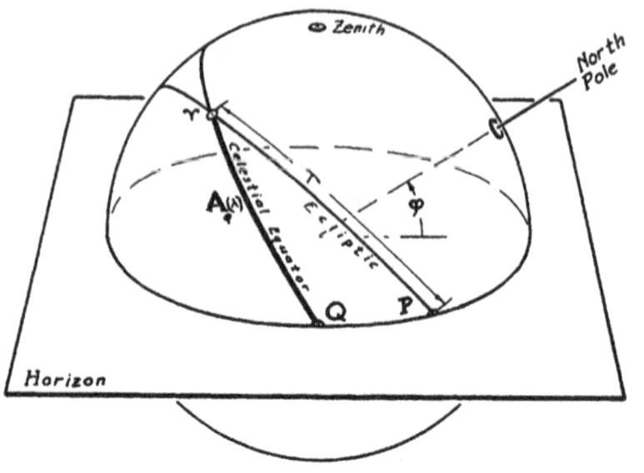

FIG. 2

astronomical in the sense that astronomical parameters are embedded in them. Those which are standard to the zījes are defined herewith:

1. The *first declination* (δ_1, *al-mail al-awwal*), or simply the *declination* (δ, *al-mail*) of an arbitrary point P (fig. 1), on the ecliptic, having *celestial longitude* (*ṭūl*) λ is the distance from P to the celestial equator. Clearly, δ depends not only on λ, but also on the value adopted for ϵ, the *obliquity of the ecliptic* (*ghāyat al-mail*, *al-mail al-a'ẓam*, Persian: *mail-i kullī*).

2. The *second declination* (δ_2, *al-mail ath-thānī*) of P is the great circle arc indicated as such on figure 1, issuing from P as before, but now perpendicular to the ecliptic.

3. The *ascensions* or *rising times* (*maṭāli' al-burūj*) are conveniently defined by use of figure 2. For an observer whose station has a (*terrestrial*) *latitude* (*'arḍ*) of φ and for an arbitrary point P of the ecliptic, having (*celestial*) longitude λ, consider the instant when P is on the local eastern horizon, i.e., the instant when it is rising. Then $A_\varphi(\lambda)$ is the distance from the vernal equinoctial point (♈) to the simultaneously rising equatorial point Q.

For the important special case where $\varphi = 0°$, i.e. when the observer is on the terrestrial equator, the risings are said to be in the *right sphere* (*al-falak al-mustaqīm*, *sphera recta*). The usage has survived in modern astronomy as *right ascension*, for which the standard symbol is α.

For a discussion of the properties of the rising times and some of their manifold applications, see O. Neugebauer, "On Some Astronomical Papyri . . .," *Trans. Amer. Philos. Soc.*, N.S., vol. 32 (1942), pp. 251–263. The notion of rising times has played a fundamental role in the history of astronomy from Babylonian times on.

In figure 2, P is the *ascendant* (*aṭ-ṭāla'*, *horoscope*, λ_H), a point of prime importance in astrology.

Closely related to the A-functions, are the problems of length of daylight (D), maximum length of daylight (max D), and the length of the unequal hours, ($\tilde{h}$). It is clear that D is a function both of φ and of the season of the year, max D increasing from twelve hours at the equator ($\varphi = 0°$) to twenty-four at the arctic circles ($\varphi = 90° - \epsilon$), where the problem degenerates. The *equation of daylight* ($\Delta D \equiv D - 12$, *ta'dīl an-nahār*) is tabulated in various ways in a number of zījes.

It was customary to divide the time from dawn to sunset (or sunset to dawn) into twelve equal parts, the *unequal hours* (*as-sā'āt az-zamāniya*, $\tilde{h} \equiv D/12$), as distinguished from the *equal hours* (*as-sā'āt al-mustawiya*, or *al-mu'tadila*). Many zījes have tables of $\tilde{h}$ as a function of the season, for a convenient fixed φ.

A technical term defined in terms of the maximum length of daylight is that of a *climate* (*iqlīm*). In Islamic astronomy the *first climate* is that portion of the northern hemisphere in which $12\frac{3}{4} \leq \max D \leq 13\frac{1}{4}$. For the *second climate* the condition is $13\frac{1}{4} \leq \max D \leq 13\frac{3}{4}$, and so on in bands of a half-hour's advance in daylight length. For a detailed presentation, see Honigmann, E., "Die sieben Klimata . . .," Heidelberg, 1929, and the above mentioned paper by O. Neugebauer.

D. EQUATION OF TIME

The right ascension of the true sun is not a precise measure of the time which has elapsed since, say, the last vernal equinox, for it is subject to two irregularities. For one thing, the true sun does not move along the celestial equator, which measures right ascensions, but along the ecliptic, and a point moving with constant speed on the ecliptic has a projection on the equator which does not move with constant speed. And secondly, the true sun does not move with constant speed even in the ecliptic, but with a speed which is maximum at perigee and minimum at apogee. Thus the difference between mean and apparent solar time, the *equation of time* (*ta'dīl al-ayām bi laiālaihā*, $E(\lambda_s)$), is the resultant of two sinusoidal components, the one due to the obliquity of the ecliptic having a period of half a year; the one due to the eccentricity having an annual period.

A table of E is standard for the zījes.

E. MEAN MOTIONS

Consider (in fig. 3) a fixed point O, the center of the earth, with a point D rotating about it, in the plane of the ecliptic and with constant speed. About D, the epicycle center, a planet P rotates, likewise with a constant speed but which is in general different from the first. The resulting configuration pictures the two fundamental mean motions associated with any planet. The angle $\bar{\lambda}$ made by the rotating radius OD and any fixed reference line through O is the planet's *mean longitude* (*wasaṭ*). It is a linear function of time, i.e.

$$\bar{\lambda}(t) = \bar{\lambda}_0 + \bar{\lambda}'t,$$

where $\bar{\lambda}'$ is a constant, the rate of change of $\bar{\lambda}$ with respect to time, the mean angular velocity of the planet. In the same manner, a the planet's *mean anomaly* (*al-khāṣat al-wusṭā*) is also a linear function of time.

Thus the constants $\bar{\lambda}_0$, $\bar{\lambda}'$, a_0, and a' are four fundamental parameters for any planet, and they are tabulated in all the zījes.

In the Ptolemaic theory the large circle of figure 3, the *deferent*, is eccentric with respect to the center of the earth, and the point on the deferent at maximum distance from O is called the *deferent apogee* (*auj*, whence the medieval Latin *aux*), denoted in the sequel by the subscript ap. The longitude of the apogee likewise is a function of time which, in the Muslim theory, may or may not be taken as equal to the motion of *precession*, the slow westward sliding of the equinoxes along the ecliptic. There are customarily tables of these motions also.

In the Ptolemaic lunar theory the *double elongation* (*al-bu'd al-muḍā'if*, $2(\bar{\lambda}_m - \bar{\lambda}_s)$) is needed for computing the true longitude, λ_m, and some zījes have tables for it. The longitude of the *ascending node* (*ra's al-jauzahar*, *caput draconis*, ☊) is also needed for the determination of the lunar latitude and was customarily tabulated with the other mean motions.

Planetary nodes are fixed with respect to the apogees, hence there is no need to tabulate their motion.

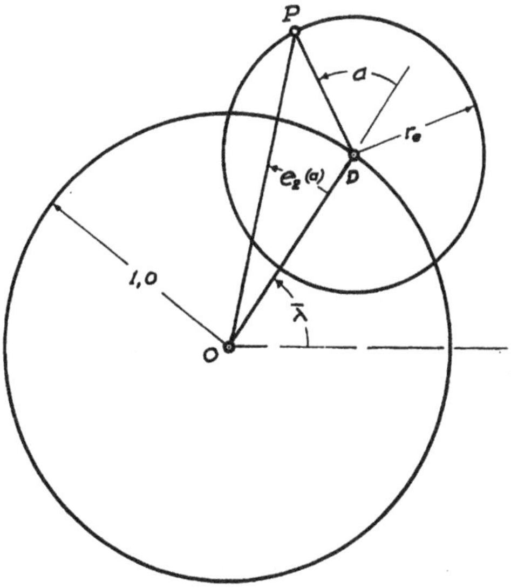

FIG. 3

In most examples the tables of mean motion have been carried to two or three fractional places only. The basic parameters from which these were computed are, however, numbers involving many more sexagesimal places. These are of interest, not that they really imply the fantastic precision of physical measurement which is indicated by the number of digits, but the occurrence of two such parameters in different documents proves influence of one on the other beyond any reasonable doubt. Therefore, when the basic parameters appear in the zīj, and have not previously been published, we insert them in the abstract. These parameters can also be "squeezed" [7] out of the ordinary tables, but considerable routine computation is involved.

F. PLANETARY EQUATIONS

With the single exception of **21** (abstracted in §6 below) all twelve zījes adopt Ptolemy's planetary model together with the corresponding numerical procedures for computing the corrections to be applied to the mean positions. It will therefore suffice to sketch it once for all here, noting in the individual abstracts only such variants of presentation as may have been observed.

Since it has no epicycle, the sun exhibits only one inequality in its motion, that due to the eccentricity of its orbit, and this is computed and tabulated in a straightforward manner on the basis of the theory. The moon and the five planets, however, have their mean motion disturbed by two independent equations, one (*the first equation*, *at-ta'dīl al-awwal*, e_1) due to the eccentric position of the epicycle-center on the deferent, the other (the *second*, *at-ta'dīl ath-thānī*, e_2) due to the position of the planet itself on the periphery of the epicycle. Instead of computing the composite equation directly on the basis of the theory and for independent variation of the two variables $\bar{\lambda}$ and a, Ptolemy simplified the computation somewhat as follows, and with negligible loss of precision.

He computed a table giving $e_1(\lambda)$ as the sum of corresponding elements in two adjoining columns and disregarding the epicycle completely. He also computed a table of e_2 for general values of a, but with the epicycle center at mean distance from the deferent center. With this are associated columns of corrections which, when added algebraically to the e_2 just mentioned, yield values of e_2 for general a but with the epicycle center at greatest (apogee) and least (perigee) distances. Finally, he tabulated also a non-linear interpolation scheme to give approximate modifications in e_2 due to general locations on the deferent. The true longitude of the planet is then

$$\lambda(\bar{\lambda}, a) = \bar{\lambda} + e_1(\bar{\lambda}) + e_2(a, \bar{\lambda}).$$

[7] This expressive usage, as well as the technique it denotes, is due to O. Neugebauer.

G. PLANETARY LATITUDES

The most spectacular aspect of the planetary motions is their characteristic apparent looped paths among the fixed stars, the retrogradations recurring periodically, but never with precisely the same loops. It is a prime requirement of any planetary theory that it produce this phenomenon; hence the model must be so set up that it pulls the planet north and south of the ecliptic, i.e. it must have a generally non-zero latitude. Moreover, to correspond with observation, maximum latitude should occur when the planet is nearest the earth, i.e. at or near epicycle perigee. Such a theory was worked out by Ptolemy, although his scheme was not the first attempt.

Somewhat as is the case with *F* above, all zījes examined save **21** (§6) and **51** (§5) have taken over the Ptolemaic latitude theory intact, except for unessential variants in presentation or parameters. (For a concise description, see *Kennedy* 2.) For the superior planets Ptolemy defined two latitude components, β_1 and β_2; for the inferior planets he added a third, β_3. All are produced by tilting of the deferent and epicycle with respect to the ecliptic plane, and the latitude of the planet is the algebraic sum of the components.

Many of the Ptolemaic latitude parameters are noted in §17 below.

For Ptolemy and for all the zījes observed the lunar latitude is given by the expression

$$\beta_m = \max \beta_m \sin (\lambda_m - \lambda_n)$$

where λ_n is the longitude of the ascending node.

The Ptolemaic maximum is 5° and is used by most of the zījes. Other maxima will be noted in individual abstracts.

H. PLANETARY STATIONS AND RETROGRADATIONS

In generating the loops referred to in *G* above an instant arrives when the forward (eastward) motion of a planet ceases, i.e. $\lambda' = 0$. The planet is then said to be in *first* or *retrograde station* (*al-maqām al-awwal* or *ar-rij'a*). It then becomes *retrograde* (*rāji'*), and after a time again stops, at *second* or *direct station* (*al-maqām ath-thānī*, or *al-istiqāma*) and resumes its direct motion. It is then said to be *muqīm*.

Ptolemy applies a theorem of Apollonius (*Almagest*, XII, 1) to compute the approximate location of the stations when the epicycle is at maximum, mean, and minimum distances from the center of the universe. By use of an interpolation scheme analogous to that applied for the planetary equations he then works out a table of positions of the stations in terms of the anomaly a for a set of positions of the epicycle intermediate between the apogee and perigee of the deferent. Most zījes likewise carry such tables, the entries being identical with or only slightly different from the Ptolemaic values.

I. PLANETARY SECTORS

Three of the zījes abstracted (in §§13, 15, and 16) have tables of *niṭāqāt* (singular, *niṭāq*), which we will call *sectors*. These are determined by four points each, either on the deferent or the epicycle. The first point is the apogee, the second a point between apogee and perigee and eastward from the former. The third point is at perigee, and the fourth eastward from the third. The precise location of the second and fourth points depends on whether the sector in question is reckoned *according to distance* (*ḥasab al-bu'd*) or *according to motion* (*ḥasab al-ḥaraka*). Roughly speaking, the second and fourth points are those at which the moving object, the planet or its epicycle center, is at mean distance from the center of the universe, in the case of the *distance sectors*. For the *motion* (or *velocity*) *sectors* the second and fourth points are those at which the object attains its mean angular velocity as viewed from the center of the universe.

In all cases, the first sector is the region between the first and second points, the second sector is between the second and third, and so on. There are thus four categories of sectors, being the combinations of *epicycle* or *deferent*, with *distance* or *velocity*.

The foregoing sketch leaves out numerous special arrangements and inconsistencies in the actual computations, as with the oval deferent of Mercury, for instance. The present writer hopes eventually to publish details in a critical edition of *Princeton, Pers.*, Ms. (Garrett) 75 currently under preparation.

The notion of sectors does not appear until the Islamic period, and only late zījes have tables of them. *Ibn Hibintā*, early in the period devotes considerable space to them in his astrological treatise, and *Bīrūnī* (*Tafhīm*, p. 110), in his introduction to astrology, defines and tabulates them, whence it may be inferred that astrology motivated their introduction, for they have no immediate application in practical astronomy.

J. PARALLAX

When a celestial object is viewed from the earth's surface its apparent position in the celestial sphere is slightly different from that computed on a strictly geocentric basis, as though the observer were at the center of the earth. The difference is known as *parallax* (*ikhtilāf al-manẓar*), and it varies inversely with the distance of the object. Parallax is always zero in the zenith, reaching its maximum on the local horizon. The moon is sufficiently close to the earth to make the problem of determining its parallax of practical importance for the computation of solar eclipses.

The *Almagest* (V, 18) contains tables of solar and lunar parallax in the altitude circle. These are reproduced in some zījes. The difference between lunar and solar parallax at a given conjunction is called the *adjusted lunar parallax* (in Persian *ikhtilāf-i manẓar-i mu'addal-i qamar*). The *Handy Tables* of Theon have a set of tables of the latter, broken up into its latitude and longitude components (P_β and P_λ). There is one table for each climate, and for each table there are two arguments: (1) integer hours of the conjunction's occurrence before or after the local meridian, and (2) initial points of the zodiacal signs, locating the conjunctions in the ecliptic.

Most of the zījes reproduce these Theon tables without change, although some later ones display tables of the same type computed for other geographical latitudes.

In the sequel the symbol h is used for altitude, zenith distance then being $90°-h$.

K. ECLIPSE TABLES

While the contents of these sections are by no means standard as between different zījes, it is possible to list the types of tables frequently occurring. They are:

Tables of $\lambda_s'(\bar{\lambda}_s)$ and $\lambda_m'(a_m)$, solar and lunar rates (at syzygies) as functions of the mean solar distance from apogee and the lunar anomaly respectively.

Tables of r_s, r_m, and r_w, apparent solar, lunar, and shadow radii respectively, as functions of $\bar{\lambda}_s$ and a_m or of the λ's.

Eclipse magnitudes are measured in *digits* (*aṣābi'*, singular *aṣba'*) defined as follows: Consider the line of centers of the luminous and eclipsing disks. The segment of this line which is common to both disks measured in twelfths of the luminous disk's diameter is the magnitude of the eclipse. Thus a total eclipse has a magnitude of twelve digits. *Areal digits* (*aṣābi' al-jirm*) are analogously defined, but the eclipsed *area* is measured in twelfths of the whole (lunar or solar) luminous area. The *Almagest* table for converting between diametral and areal digits is frequently reproduced.

Tables of mean conjunctions and oppositions are usually in terms of the Hijra calendar, since it is lunar. Functions tabulated are t, the time of day of the mean conjunctions or opposition in question, a_m, λ, and λ_{apS} at the same time.

There are usually tables of lunar eclipses and of solar eclipses, of one of two types. One arrangement is to set up a table with the moon assumed at epicyclic apogee, another with the moon at epicyclic perigee, and an interpolation arrangement to take care of intermediate values of the anomaly. Eclipse magnitude and immersion time are tabulated against a range of β_m.

The alternative is to use a two-argument table in which magnitude and duration are tabulated as functions of β_m and λ_m'.

The *inclination* (*inḥirāf*) of an eclipse is the angle between the ecliptic and the line joining the centers of the lunar and shadow (or solar) disks, at the instant of first contact. If the eclipse is total, the same angle may be computed for internal tangency of the two disks.

L. VISIBILITY TABLES

The very difficult problem of predicting the date of first visibility of the lunar crescent as well as the apparitions and disappearances of the planets was of central importance in motivating Babylonian astronomy. In the *Almagest*, however, visibility conditions receive very little consideration. The very last tables of the work give sets of limiting planetary elongations which when exceeded insure visibility of the planet in question. These values, quite properly, follow as corollaries of the general planetary theory.

The *Almagest* table is for the latitude of Babylon only. There is an entry each for morning risings and evening settings of the superior planets; for the inferior planets there are additional columns for evening risings and morning settings, all the above tabulated for each zodiacal sign, to two places. The *Almagest* has no material at all on first visibility of the lunar crescent.

The rise of Islam saw a recrudescence of interest in visibility problems, partly because the Hijra calendar is strictly lunar, and the fasting month of Ramaḍān, in particular, begins in a given locality only when the new crescent is actually seen by a competent authority.[8]

As will be noted in the individual abstracts, many zījes contain extensive visibility tables, practically none of which have been touched by modern scholarship. The only published material on medieval visibility problems known to this writer is: Neugebauer, O., "The Astronomy of Maimonides and its Sources," Hebrew Union College Annual, vol. XXII, pp. 322–363, and *Nallino, Batt.*, vol. I, pp. 266-272, vol. II, pp. 255–269.

M. GEOGRAPHICAL TABLES

Most zījes have extensive lists of cities and other geographical localities, together with their terrestrial latitudes and longitudes.

N. STAR TABLES

Practically all zījes have tables of positions of the fixed stars, usually reported in ecliptic coordinates sometimes in equatorial coordinates also. Star magnitudes are frequently given.

It was customary to assign to each fixed star a quality or *temperament* (*mizāj*) resembling one of the planets. These are often indicated in the star tables with the terminal letter of the associated planet. (*Cf. Bouché-Leclercq*, p. 132; *Tetrab.*, I., 8.)

O. ASTROLOGICAL TABLES

It will be useful at this point to introduce a number of concepts which, while essentially astronomical in character, were defined and applied for astrological purposes. For detailed information the reader is referred to *Tetrab.* and to *Bīrūnī, Tafhīm*.

At any given instant the four *centers* (*pivots cardines*; Arabic *al-autād*, singular *al-watad*) on the ecliptic are: the *ascendant* defined on p. 140 lower *culmination* (*imum coeli*, Arabic *ar-rābi'*, the fourth, since it is the initial point of the fourth *house* defined below), the *descendant* or *setting point* (*al-ghārib* or *as-sābi'*, the seventh) opposite the ascendant, and *upper culmination* (*al-'āshir*, the tenth, or *wasaṭ as-samā'*, but the latter term is ambiguous; see *Kennedy*, 1, §3. The unequal four arcs into which these points divide the ecliptic are each in their turn divided into three arcs, each of the resulting twelve being called a *house* (*beit*, pl. *buyūt*) numbered from the ascendant opposite to the direction of the daily rotation.

The arcs defining the houses are frequently found by trisecting the four major arcs. There are other methods of effecting the division, and in any case the process, called the *equalization of the houses* (*taswiyat al-buyūt*) is frequently tabulated in the zījes.

It was a widely held view that any given planet exerted an influence on other planets in the half of the ecliptic following it, by casting back rays, missiles as it were, upon them. The determination of the precise point of impact, called the *projection of the rays* (*maṭāriḥ ash-shu'ā'āt*), need not have been a complicated process. But, as was the case with the equalization of the houses, there were alternative methods of effecting the transformation of the ecliptic into itself which was involved, utilizing the local horizon, the celestial equator, and so on.

The instant at which the sun crosses the vernal equinoctial point is called *year-transfer* (*taḥwīl as-sina*) and was considered to be of great significance. Even to the present in Iran the situation of the individual at the moment of transfer is supposed to affect his destiny through the coming year. In the same manner *nativity-transfers* (*taḥāwīl sinī al-mawālīd*) are the instants, in successive years, at which the sun enters the same point of the zodiac in which it stood when the individual was born. These two concepts evidently occupied a central position in the practising astrologer's stock in trade, for the *Fihrist* notes no less than fourteen books with one of these two terms as titles as having been written in early Abbasid times.

Closely allied to the above notions is that of the *excess of revolution* (*faḍl ad-daur*, in medieval Latin *revolutio anni*), which may be defined as the time measured in degrees of the daily rotation ($360° = 24^h$) by which the solar year exceeds the Persian (i.e., Egyptian) year of 365 days. Several zījes have tables of this quantity, in some cases the year being the tropical, sometimes the sidereal. Given the excess for a year, multiplication by 1/6, 0 = 0; 0, 10 and addition of the result to 6, 5 gives the corresponding solar year in days.

[8] *Bīrūnī, Chron.*, p. 34.

Astrological predictions regarding the length and vicissitudes of a given individual's life were based on the association of a moving point on the ecliptic with the life in question. Each degree of the point's motion corresponds to a year of the person's life. The starting place, initial impetus, and eventual stoppage of the point (the *aphesis*, *directio*, Arabic *tasyīr*) are governed by complicated considerations connected with the horoscope. Other indicators, moving at different rates, but all having to do with the nativities of persons, are the *progressions* (*intihā'āt*), and *firdaria* (*fardārāt*).

The same type of indicators, progressing far more slowly (hence with vastly greater periods), were thought to hold sway over events in the world as a whole, as though the existence of the universe were likened to the life of a single person. (See *Tetrab.*, III, 10; *Bīrūnī*, *Tafhīm*, pp. 239, 255, 321–326; *Nallino*, *Batt.*, vol. I, p. 325, vol. II, p. 339, 354; Encycl. of Islam, vol. IV, p. 694.)

Several zījes (e.g., **14**, **27**, and **54**) have tables of the motion of a celestial object called *al-Kaid*, described as one of the stars having a tail, evidently some variety of comet. CCAG V, 3 p. 147, 16 mentions these tables which concern the motion of "Καϊτ, a star considered maleficent by the *Hindus*." A footnote on the same page says "Kaïd . . . seems to be nothing but the descending lunar node" (quoting Gildemeister). Obviously it is assumed Kaïd = Kētu which is the Sanscrit name for the descending node. This is disproved by the period of 144 years. Al-Khāzinī (**27**, f. 129) gives a rule for determining its position in the ecliptic. It is, in effect,

$$\lambda = \frac{360}{144}\,(\text{residue modulo 144 of } (y + 54))$$

where y is the Yazdigird year for which the position is desired and λ the position in degrees. Thus the motion has a period of 144 years, traversing a sign in twelve years, with epoch taken as fifty-four years before the Yazdigird era. *Ibn Hibintā* (f. 75) devotes a paragraph to al-Kaid following a general discussion of comets.

P. MISCELLANEOUS

In this category tables and topics are placed which fall naturally into none of the above.

5. ABSTRACT OF 51, THE PURPORTED *AZ-ZĪJ AL-MUMTAḤAN* OF *YAḤYĀ IBN ABĪ MANṢŪR*, *c.* 810

It should be stated again that material in this manuscript is to be assumed *not* a part of the original zīj of Yaḥyā unless there is reason for thinking otherwise. The introduction is probably genuine as well as the title on f. 3 "The First Chapter, On the Division of the Arabic Months." From here on there are no chapter headings as such; there are some self-contained *risā'il*, and a few tables specifically ascribed to Yaḥyā.

A. CHRONOLOGY

There is material, not necessarily complete, on the Hijra, Seleucid, Yazdigird, Coptic, and Jewish calendars.

B. TRIGONOMETRIC FUNCTIONS

There are three-place tables of:

$$\text{Sin}\,\theta, \quad \text{for} \quad \theta = 0;0°, 0;1°, 0;2°, \cdots, 5;0°,$$

(This table is from Abū al-Wafā' (**73**).)

$$\text{Sin}\,\theta, \quad \text{for} \quad \theta = 0°, 1°, 2°, \cdots, 90°,$$

with tabular differences. (This one is from Kūshyār (**7**).)

$$\text{Tan}\,\theta, \quad \text{for} \quad \theta = 0°, 1°, 2°, \cdots, 89°.$$
$$12\cot\theta, \quad \text{for} \quad \theta = 1°, 2°, 3°, \cdots, 90°.$$

C. SPHERICAL ASTRONOMY

In this category are tables of $\delta(\theta)$, to seconds of arc, for each integer degree of θ, and for $\epsilon = 23;33°$.

$A_\varphi(\lambda)$ to minutes of arc, for each integer degree of λ, for:

$\varphi = 0°$, with tabular differences,
$\varphi = 0°$, without tabular differences,
$\varphi = 36;0°$ (the latitude of Raqqa, where the observations of al-Battānī (**55**) were made)
φ of Baghdād,
φ of Mosul.

There are tables, computed to minutes of arc for each integer degree of the argument, for

$$D(\lambda_s) \quad \text{for} \quad \varphi = 36;0°,$$

and

$\bar{h}(\lambda_s)$, for $\varphi = 36;0°$ and for the latitude of Mosul.

There is a table, computed to minutes of arc, for each integer degree of the argument, of

$$\max h_s(\lambda_s), \quad \text{for} \quad \varphi = 36°.$$

Two tables, ascribed to Abū al-Wafā' (**73**), display

$$n\cdot\text{Sin}\,\epsilon \quad \text{and} \quad n\cdot\text{Cos}\,\epsilon, \quad \text{for} \quad n = 1, 2, 3, \cdots, 60.$$

A function widely used in Islamic astronomy, but rarely tabulated, is $\bar{h}_e$, the *latitude of the visible climate* (*'arḍ iqlīm li'r-rū'ya*, see *Kennedy*, *1*, §3). For a given time and a given terrestrial location, it is the complement of the angle the ecliptic makes with the local horizon. This zīj has a table of $\bar{h}_e(\lambda_H)$, to seconds of arc for each integer degree of the argument, for $\varphi = 33;21°$ (Baghdād).

D. EQUATION OF TIME

There is a table, computed to seconds, of $E(\lambda_s)$, for each integer degree of the argument.

E. MEAN MOTIONS

These tables are all to three fractional places and locate all of the means mentioned in the outline at the beginnings of Yazdigird years 0, 20, 40, · · ·, 600. The motion of each of the means is given for 1, 2, 3, · · ·, 20 "Persian" years (i.e. years of 365 days), for 1, 2, 3, · · ·, 12 Persian (i.e. 30-day) months, for 1, 2, 3, · · ·, 29 days, and for hours and minutes. The base values, from which the Mumtaḥan tables were computed, are given by Ibn Yūnis (**14**, *Caussin*, pp. 230–237). The solar motion at least, of this copy, is that given for the Mumtaḥan.

Precessional motion is tabulated to three fractional places for Hijra years 1, 2, 3, · · ·, 30, 31, 61, 91, · · ·, 631. The parameter is 0; 0, 54, 44, 20° per Hijra year, the same as that of **16**, as nearly as can be seen from this approximate determination.

F. PLANETARY EQUATIONS

These tables are of the standard form, computed for each degree of all arguments. e_s is given to seconds of arc, the equations of the moon and planets to minutes. The title of the table giving the equations of Mars has the phrase ". . . the Observation of . . . Ibn al-A'lam." To this has been added, in a different hand, and above the line, "and of Yaḥyā bin abī Manṣūr."

G. PLANETARY LATITUDES

This zīj has tables and explanation for an extraordinarily primitive planetary latitude theory. In all cases the latitude is given by an expression of the form

$$\beta(\bar{\lambda}) = \max \beta \cdot \sin(\bar{\lambda}),$$

where here $\bar{\lambda}$ is the planet's longitude, measured from the ascending node. For the superior planets, however, different max β are taken depending on whether

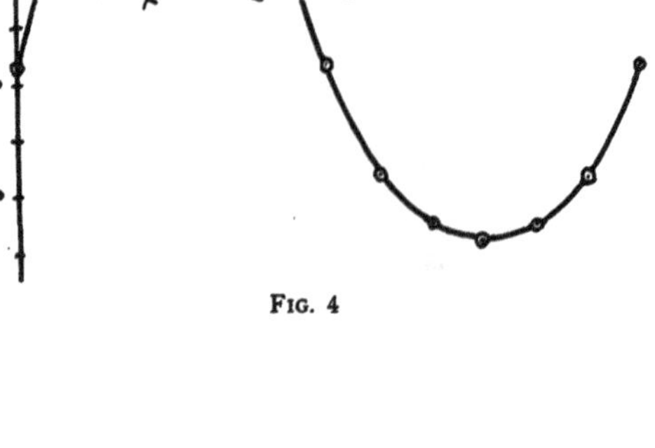

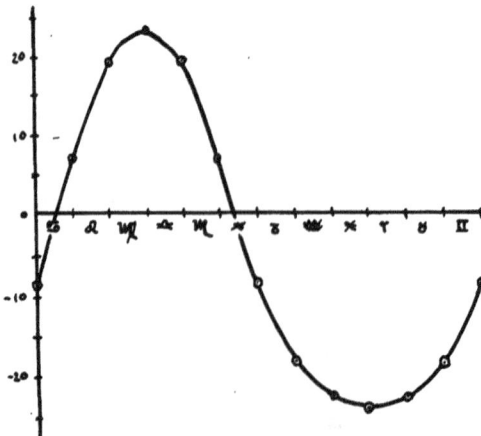

FIG. 4

the latitude is north or south. Except for an unimportant variant, the same maxima are given by *Ibn Hibintā* among other sets of parameters he ascribes either to the Shāh Zīj (**30**) or to the Sindhind (**28**). The maximum latitudes, however, are not specifically assigned to either. These parameters are displayed in §17 which also discusses the possibilities as to the origins of the theory.

For this zīj max β_m is 4; 30° as with **21** (§6, G below).

There is a table of the longitudes of the planetary nodes, the values of which in general coincide with those of Khwārizmī (**21**) and az-Zarqālī (**24**).

H. STATIONS AND RETROGRADATIONS

There is a table of stations computed to minutes of arc for each degree of the epicycle center's distance from deferent apogee.

I. SECTORS

None.

J. PARALLAX

This zīj has the same table of P_β found in Khwārizmī's zīj (**21**, §6, J), which see, but with no accompanying explanatory text, and with the title, "A Table of Lunar Latitude Difference for Visibility."

K. ECLIPSE THEORY

There is a table of solar eclipses with two arguments: eclipse magnitude = 0; 15, 0; 30, 0; 45, 1; 20, 1; 40, · · ·, 12; 20 digits and the corresponding β to seconds, and λ_m' = 0; 27, 30°, 0; 28, 30°, 0; 29, 30° · · ·, 0; 33, 30° per hour.

The entries then give the time of immersion to minutes.

There is a "Table of Azimuth (? *samt*) for Knowledge of the Solar Eclipse," computed to minutes of arc for each degree of the ecliptic. Its graph is sketched as figure 4, and has the appearance of a sine wave pinched in horizontally at one half and widened correspondingly in the other. The application of this table is unknown to the present writer.

L. VISIBILITY THEORY

The only visibility material, except perhaps the table noted in *J* above, is a lunar ripeness table resembling, but not identical with, that of Khwārizmī described in §6, L below. It gives two-place entries for each sign.

M. GEOGRAPHY

None.

N. STAR TABLES

The positions of twenty-four stars are given, both in ecliptic (λ, β) and equatorial (α, δ) coordinates, to minutes. These are stated to have been obtained by observations made in 214 A.H. (A.D. 829/830) at Damascus and at the Shamāsī observatory in Baghdād. The coordinates are the same as those given in **16** and evidently are from the original Mumtaḥan Zīj. The table is excerpted in §17 below.

There is another table, for eighteen stars only, for the year 380 (≈A.D. 1012) of Yazdigird. Presumably it has been computed from the first, for the latitudes are the same and a precession correction has been added to the λ's. For each star the equatorial coordinates also are given, as well as half the daylight arc, the sine of the daylight arc, the meridian altitude, the sine of the latter, the degree of the simultaneously rising point of the ecliptic and its ascension.

O. ASTROLOGY

There is a table of planetary periods and orbs, another of zodiacal indicators, and another giving the triplicities, terms, decans, and exaltations for each sign.

A table of the *excess of revolution* is given, computed to seconds, for 1, 2, 3, ···, 10, 20, 30, ···, 90 years. For one year it is 1, 26; 43, 39°, to which corresponds a tropical year of 6, 5; 14, 27, 16, 30^d. Both values are independently attested from **15** (§7, O) as being the results of the Mumtaḥan observations at the Shamāsī Observatory in Baghdād. Hence this is a secure (rounded off) value for the Mumtaḥan tropical year, or at least for the results of one observation.

On f. 8 appears a set of values of the excess of revolution as used by various astronomers. These are transcribed below, followed by the length of (sidereal or tropical) year implicit in each.

				Zīj *Serial No.*	*Suter No.*
1.	Māsha'allāh	93; 15, 0°	6, 5; 15, 32, 30^d		8
2.	aṭ-Ṭabarī	93; 9, 40	6, 5; 15, 33, 36, 40	**65**	
3.	Shāh Zīj	93; 15, 0	6, 5; 15, 32, 30	**30**	
4.	Ptolemy	88; 40, 0	6, 5; 14, 46, 40		
5.	The Sindhind	93; 0, 15	6, 5; 15, 30, 2, 30	**28**	
6.	Al-Ḥasan bin Sahl	93; 15, 0	6, 5; 15, 32, 30		27
7.	Al-Khwārizmī	93; 2, 0	6, 5; 15, 30, 20	**21**	
8.	Zīj al-Hazārāt	93; 14, 0	6, 5; 15, 32, 20	**63**	
9.	The Mumtaḥan	86; 35, 55	6, 5; 14, 25, 59, 10	**51**	
10.	Ḥabash	86; 42, 17	6, 5; 14, 27, 2, 50	**15**	
11.	An-Nairīzī	86; 36, 0	6, 5; 14, 26	**46**	
12.	Al-Battānī	86; 36, 0	6, 5; 14, 26	**55**	

1, 3, and 6 above all have the same value. This parameter is mentioned by *Hāshimī*, who says it was used by the Persians and by Māsha'allāh. *Bīrūnī* (*Chron.*, p. 121) says it was used by the Persians. Since Ibn Sahl is known to have been a Persian, and Māsha'allāh, though Jewish, was strongly under Persian influence, all the information is consistent, and this parameter can be accepted as well attested.

The value of 2 is given by *Hāshimī*, but without attribution to an individual.

The statement of 4 is incorrect. Ptolemy's tropical year is in fact 6, 5; 14, 48^d, to which corresponds an excess of revolution of 88; 48°.

The value for 5 does not yield an attested Hindu length of year. If the excess of revolution is restored as 93; [2], 15° the corresponding year becomes 6, 5; 15, 30, [2]2, 30, which is that of the Brahmasiddhānta, the Siddhānta-Śiromaṇi (*Nallino, Batt.*, vol. I, p. 206), and the Arabic Sindhind (**28**) according to Suter (*Khwar.*, p. 103), and *Hāshimī*.

In *Suter, Khwar.* three lengths of year are to be found, 6, 5;15, 32, 30 (p. 42), 6, 5; 15, 32, 35 (p. 65), and 6, 5; 15, 30, 22, 30 (p. 230). None of these is the same as that implied by our text in 7 above.

The entries for 8, 9, and 10 have not been seen elsewhere by this writer. Moreover, in the case of 9, it conflicts with the attested value above, from the same manuscript.

The common entry for 11 and 12 is indeed the tropical year adopted by al-Battānī (*Nallino, Batt.*, vol. I, p. 42). According to *Millás Vallicrosa* (p. 76), it was commonly used by other Arab astronomers, including Thābit bin Qurra (**93**). Ibn Yūnis (*Caussin*, p. 74) says that an-Nairīzī used the value 86; 35, 12°, whereas had he computed it from his own (Nairīzī's) length of year it would have been 86; 43, 13, 18°. Neither is the value given in 11 above.

P. MISCELLANEOUS

There appears in two different places (ff. 10 and 70) a table with the strange title of the *solar latitude* (*'arḍ ash-shams*) having approximately the form 0; 48, 32 sin θ. An application of this table is described in *Kennedy, 1*, §9.

There is a table of the function 24 d/λ' for

$$d = 1', 2', 3', \cdots, 60',$$
$$\lambda' = 57', 58', 59', \cdots, 62' \text{ per day},$$

computed to four places. For the application of such a table, see §13, P below.

6. ABSTRACT OF 21, THE ZĪJ OF *AL-KHWĀRIZMĪ*, *c.* 840

(Page references are to the publication, *Khwar.*)

A. CHRONOLOGY

This zīj has material on the Hijra, Seleucid and Yazdigird calendars (pp. 109–114).

B. TRIGONOMETRIC FUNCTIONS

There are tables (pp. 169–170, 174) of Sin θ, to three places, for $\theta = 0°, 1°, 2°, \cdots, 360°$, and 12 cot θ, to one fractional place, for $\theta = 0°, 1°, 2°, \cdots, 90°$.

C. SPHERICAL ASTRONOMY

There are also tables of $\delta(\theta)$, to seconds of arc, for each integral degree of θ, for $\epsilon = 23;51°$ (pp. 132–137).

$A_0(\lambda) + 90°$ to seconds of arc for each integer degree of λ, i.e. for *sphera recta* only (pp. 171–173). *Bīrūnī* (in *Risā'il*, II, p. 129) states that this zīj contains a table called "Differences of Ascensions for the Earth," an arrangement for modifying right ascensions in order to obtain oblique ascensions. This table has not survived in the published version. However, a table of the same type exists in a Latin manuscript of the fifteenth century. It has been published by Neugebauer and Schmidt in "Hindu Astronomy at Newminster in 1428," *Annals of Science*, vol. 8, 1952, pp. 221–228.

D. EQUATION OF TIME

There is a table (pp. 181–182) of $E(\lambda_s)$, computed to two places, for each integer degree of the argument.

E. MEAN MOTIONS

These tables (pp. 115–131) are all computed to seconds of arc. Positions of $\bar{\lambda}_s$, $\bar{\lambda}_m$, a_m, and λ_n are given for Hijra years 0, 30, 60, $\cdots$, 720, (since a 30-year cycle is involved). The mean motions are given for 1, 2, 3, $\cdots$, 30 Hijra years; 1, 2, 3, $\cdots$, 12 Hijra months; 1, 2, 3, $\cdots$, 29 days; 1, 2, 3, $\cdots$, 24 hours, and 2, 4, 6, $\cdots$, 60 minutes. For the three superior planets, $\bar{\lambda}$ and $\bar{\lambda}'$ are tabulated, the arguments being as above except that the range of years extends to 570 A.H. only. For the inferior planets, a and a' are shown, for the same domain of the arguments as for the superior planets except that the range of days is 1, 5, 10, 15, 20, 25, 30, 60, and that of hours is 1, 3, 6, 12, and there are no columns for minutes.

According to *Ibn al-Qifṭī* (p. 326) al-Majrīṭī converted the mean motion tables to the Hijra epoch, the zīj in its original form having used the epoch of Yazdigird.

F. PLANETARY EQUATIONS

The structure of these tables (pp. 132–167), hence the underlying theory, differs so markedly from that of the Almagest that it is thought worth while to describe it in some detail.

As usual, there is but one equation for the sun. But whereas all other zījes here abstracted tabulate a solar equation computed directly on the basis of an eccentric circular orbit, Khwārizmī's solar equation, tabulated to seconds for each degree of the argument (pp. 132–137), is symmetrical about an ordinate through 90° and is very close to $2;14° \sin \theta$. Thus the true longitude of the sun is (very nearly) given by the expression

$$\lambda_s(\bar{\lambda}_s) = 2;14° \sin(\bar{\lambda}_s - \lambda_{apS}).$$

This is an example of a type of trigonometric interpolation commonly used by the medieval Hindus, but which is found already in the *Almagest* (XIII, 4) in Ptolemy's planetary latitude theory.

The case of the moon is of even more interest, for it is well known that pre-Ptolemaic lunar theory employed only one equation, and so does Khwārizmī. The single table of the lunar equation has precisely the character of that for the sun, but a different maximum. It is approximately $4;56° \sin \theta$.

The tables for the planetary equations (pp. 138–167) are computed for each degree of the arguments, to minutes of arc. The equation of the center, e_1, has in every case the now familiar sinusoidal form $\max e_1 \sin \theta$ appearing as the fourth column of the table. (The parameters are shown in §17 below with the corresponding Ptolemaic ones.) The equation of the anomaly, $e_2(a)$, is computed on the basis of an epicyclic anomaly with no deferent eccentricity as indicated in figure 3. This is the third column of the table. But the inventor of the theory apparently realized that the two equations are not independent; to make one affect the other he forced the planet's (deferent) apogee to oscillate with the period of the anomaly and an amplitude half that of the anomalistic equation. To that end, the second column of the table is

$$\lambda_{ap} - e_2(a)/2$$

This variable apogee is evidently used in the argument for the first equation. Thus the true equation of the planet is given by the expression

$$\lambda(\bar{\lambda}, a) = \bar{\lambda} + e_1(\bar{\lambda}, a) + e_2(a_1)$$
$$= \bar{\lambda} + \max e_1 \sin\left(\bar{\lambda} - \lambda_{ap} + \frac{e_2(a)}{2}\right) + e_2(a_1).$$

where $a_1 = a - e_1(\bar{\lambda}, a)$.

From the critical remarks of *Bīrūnī* (in *Risā'il*, I, pp. 131, 174) it is clear that the method of trigonometric interpolation was that actually used by Khwārizmī in the zīj in its original form.

G. PLANETARY LATITUDES

Although the physical model underlying the planetary latitude tables of this zīj (pp. 132–167) can only be conjectured, it is clear that it also was radically different from that of Ptolemy.

Two functions are tabulated for each planet, the first to minutes, the second to seconds of arc, both for each integer degree of the arguments. The first, $L_1(\theta)$, is symmetrical about the ordinate through $\theta = 180°$. It has a maximum at 0°, falling off slightly to a minimum at 180°, and has horizontal tangents at both these points.

The second is

$$L_2(\theta) = \max L_2 \sin \theta.$$

Basic parameters of L_1 and L_2 for all the planets are displayed in §17 below.

At any given instant the latitude of the planet is defined as

$$\beta(\bar{\lambda}, a) = \frac{L_2(\bar{\lambda})}{L_1(a)},$$

where $\bar{\lambda}$ is the argument of the latitude, the mean longitude measured from the ascending node.

Figure 4a shows a conjectured reconstruction of the model. A non-eccentric deferent of radius R makes a fixed angle of $\max L_2$ with the ecliptic plane as shown. The epicycle center moves along the deferent, but the epicycle plane remains continually parallel to the ecliptic plane. The planet is at P and for the general situation its latitude is the angle PDP'. In the case of the outer planets, and for properly chosen parameters, this model is in fact a first approximation to the actual situation in space. For the epicycle, maintaining itself in a plane parallel to the ecliptic, corresponds to the earth's orbit around the sun. And the tilted deferent corresponds to the planet's orbit, larger than the terrestrial one.

But for the inner planets the model is fundamentally wrong. For then the deferent corresponds to the earth's orbit and should stay in the ecliptic plane, while the epicycle, being now the planetary orbit, should displace itself around the deferent, making a slight angle with it. An approximation to this result is achieved in fact by the Ptolemaic latitude components.

Returning, however, to the Khwārizmī model, assume a special position of the epicycle, at its maximum distance from the ecliptic ($\bar{\lambda} = 90°$), and for such a position define the function $\max \beta(a)$ as shown in the figure, the single variable being the anomaly a. Then, since

$$\tan(\max L_2) = \frac{c}{R}, \quad \text{and} \quad \tan(\max \beta(a)) = \frac{c}{\rho(a)},$$

$$R \tan(\max L_2) = \rho(a) \tan(\max \beta(a)),$$

and

$$\max \beta(a) \approx \max L_2 \left(\frac{R}{\rho(a)}\right).$$

Now use trigonometric interpolation to define a latitude function for general values of $\bar{\lambda}$ putting

$$\beta(\bar{\lambda}, a) \equiv \max L_2 \left(\frac{R}{\rho(a)}\right) \sin \bar{\lambda}.$$

Moreover, if we put

$$\rho(a)/R = L_1(a), \quad \text{and} \quad \max L_2 \sin \bar{\lambda} = L_1(\bar{\lambda}),$$

then the L-functions have the character of those tabulated in the zīj.

It is possible to apply a numerical test to see how well the conjectured model fits the tables. For if (fig. 3) the epicycle radius of a certain planet is r_e, then from the expression $r_e = \text{Sin}(\max e_2)$, the epicycle radii can be computed from the tables of e_2. And if the model is valid we should have, for all planets,

$$\frac{1,0 + r_e}{1,0 - r_e} = \frac{\max L_1}{\min L_1}.$$

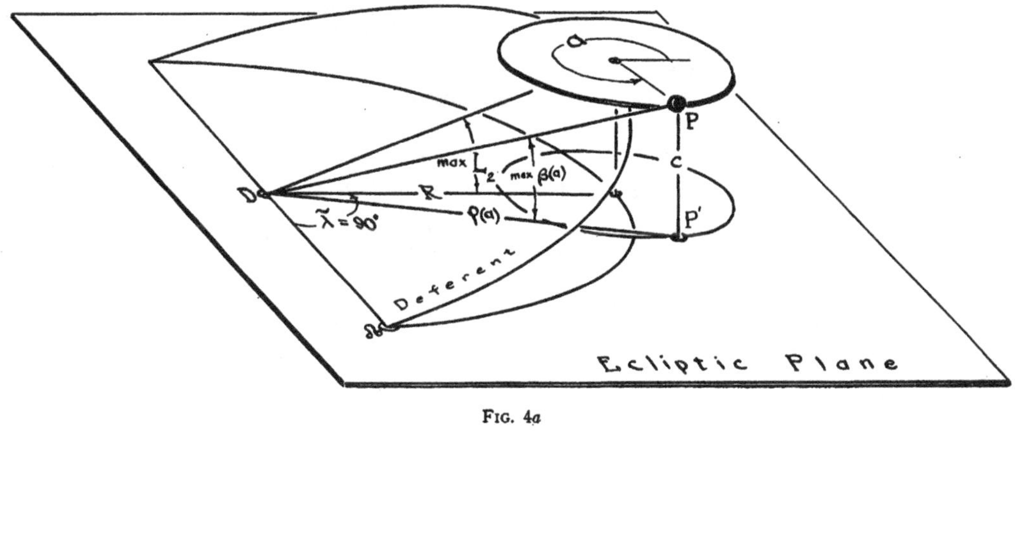

FIG. 4*a*

The table below displays the results of such a computation for the Khwārizmī Zīj.

	r_e	$\frac{1,0+r_e}{1,0-r_e}$	$\frac{\max L_1}{\min L_1}$
♄	6;0	1;13	1;20
♃	11;20	1;27	1;41
♂	39;0	4;45	5;0
♀	44;0	6;31	6;36
☿	22;0	2;9	2;11

It will be observed that the correlation is excellent for the inferior planets, deteriorating for the other planets. A conclusion is that the reconstruction is only partially successful.

For this zīj, as with **51** (§5, G), max $\beta_m = 4;30°$.

H. STATIONS AND RETROGRADATIONS

As in §5, H (pp. 138–167).

I. SECTORS

None.

J. PARALLAX THEORY

Parallax components are given by tables (pp. 191–192), computed to two fractional places for each integer degree of the arguments, of the functions

$$P_\beta = 0;48,45° \sin \bar{h}_e, \quad \text{and} \quad P_\lambda = 1;36 \sin \theta(t),$$

expressed in hours, where θ satisfies the relation $\theta(t) = t + 24 \sin \theta(t)$. See §5, C for a definition of $\bar{h}_e$. The argument t is the length of the projection on the equator of the ecliptic segment whose end points are: (1) the foot of the perpendicular dropped from the zenith to the ecliptic and (2) the position of the conjunction on the ecliptic.

The P_β table gives results which are quite accurate, the underlying theory and parameters being the same as those of the method explained in the (Hindu) *Sūrya-Siddhānta*.

Results yielded by the P_λ table are much less accurate. Although probably Hindu in origin they cannot be identified with any available theory.

See §8, J and §14, J below, and *Kennedy, 1.*

K. ECLIPSE THEORY

There is a table (pp. 175–180) giving

$\lambda_s'(\bar{\lambda}_s)$ and $\lambda_m'(a_m)$ in minutes of arc per hour, and $r_s(\bar{\lambda}_s)$ and $r_m(a_m)$ in degrees,

all computed to two places for each integer degree of the arguments. *Bīrūnī* (in *India*, transl., vol. II, p. 79) says that in this zīj Khwārizmī used the method of the Karaṇasara (**X219**) and the Khaṇḍakhādyaka (**X214**) for determining r_s and r_m.

There are tables (pp. 183–186) of mean conjunctions and oppositions giving, for Hijra years 1, 31, 61, ···, 511, and changes for 1, 2, 3, ···, 30 Hijra years and each lunar month,

$$t, \lambda, a_m, \quad \text{and} \quad (\bar{\lambda}_m - \lambda_n),$$

for the mean conjunction or opposition in question.

There are two tables of lunar eclipses (pp. 187–190), one for greatest lunar distance, the other for least. The argument runs at intervals of half a degree through the range

$$180° - 10;50° \leq |\lambda_m - \lambda_n| \leq 180° + 10;50°,$$

for the moon at greatest distance, and through the range

$$180° - 13;17° \leq |\lambda_m - \lambda_n| \leq 180° + 13;17°,$$

for the moon at least distance.

For these arguments, the following quantities are tabulated:

(1) *Eclipse magnitude*, in digits computed to minutes,
(2) *Immersion time*, to seconds,
(3) *Duration of totality*, where applicable, to seconds.

An interpolation scheme for intermediate lunar distances is also given.

In like fashion there are two solar eclipse tables (p. 193) laid out as above, except that there is no column for duration of totality, and the limits are $|\lambda_m - \lambda_n| \leq 6;37°$ and $|\lambda_m - \lambda_n| \leq 7;11°$ respectively.

The *Almagest* table for converting between diametral and areal eclipse digits is also given.

L. VISIBILITY PROBLEMS

The only visibility material in this zīj is a *lunar ripeness table* (p. 168) displaying a function $f(\lambda_m)$, computed to two places at ten degree intervals around the ecliptic. Its graph is sketched in figure 5. The method of computation of this function was unknown to the editor of the zīj and is unknown to the present writer. It is applied as follows. The evening of first visibility of the lunar crescent each month will be the first evening for which the relation

$$\lambda_m - \lambda_s + \beta_m > f(\lambda_m)$$

is satisfied.

M. GEOGRAPHY

This zīj has no geographical section, probably because Khwārizmī composed a separate treatise of this nature published by Mžik, H. as "Das Kitāb Ṣūrat al-Arḍ des abū Ğa'far Muḥ. ibn Mūsā al-Ḫuwārizmī," Leipzig, 1926.

N. STAR TABLE

There is no star table.

O. ASTROLOGICAL TABLES

There is a table (pp. 194–205) of *Equalization of the Houses*, which, for a horoscope in each degree around

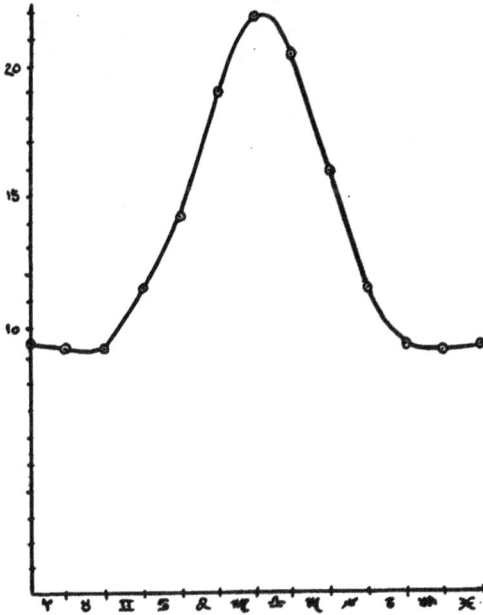

FIG. 5

the ecliptic, gives the initial points of each of the twelve astrological houses.

There is also a very extensive table (pp. 206–229) of *Projections of the Rays* (*Maṭāriḥ ash-Shu'ā'āt*) giving not only the locations of the three main aspects (sextile, quartile, and trine) as projected back from each five degree interval around the ecliptic, but the projections of all sets of 30° intervals measured from the same five-degree intervals. This table takes by far more space in the zīj than any other. It is not the work of Khwārizmī himself, but rather that of al-Majrīṭī, from whose rescension the Latin translation was made. For one thing, the φ for which the table is computed is given in the zīj as that of Cordova rather than Baghdād, Khwārizmī's location. And for another thing *Ibn Hibintā* gives a table of projections of the rays which he specifically ascribes to Khwārizmī. This table is computed for several climates, but none of the entries coincide with corresponding ones in the zīj, and the argument is for each thirty degrees rather than five.

There is a table (p. 230) of the *excess of revolution* for use in computing nativity transfers, for 1, 2, 3, $\cdots$, 10, 20, 30, $\cdots$, 100 years, the function being expressed both in days and hours (to minutes) and in degrees (to seconds).

The excess for one year is 93; 2, 15, 0°. According to *Hāshimī*, this is the value used in the Sindhind (**28**). To it corresponds a sidereal year of 6, 5; 15, 30, 22, 30^d, which is a well-established Hindu parameter. (*Cf.* §5, O above.)

The last table (p. 231) in the zīj gives for each sign the domiciles, exaltations, decans, etc.

A small table on page 115 gives, to seconds of arc, the position of the mean sun in the ecliptic at the instant when the true sun enters each of the signs.

7. ABSTRACT OF 15, THE BERLIN COPY OF A ZĪJ OF *ḤABASH AL-ḤĀSIB*, *c.* 850

Again, the reader should bear in mind that this manuscript, like that of **51** (§5) above, cannot be taken as the unaltered work of its purported author.

A. CHRONOLOGY

Text and tables are present for the Hijra, Yazdigird, Seleucid era, the era of Philip, Coptic, and Roman (era of Augustus) calendars.

B. TRIGONOMETRIC FUNCTIONS

There are three-place tables, with entries for each integer degree of θ, of:

Sin θ,
Tan θ,
Cot θ,
Vers θ,
and Csc θ.

C. SPHERICAL ASTRONOMICAL FUNCTIONS

There are also tables of:

$$\delta(\theta),\quad \text{Sin}\,\delta(\theta),\quad \text{and}\quad \text{Cos}\,\delta(\theta),$$

all computed to three places, for each integer degree of θ, putting $\epsilon = 23;35°$.

$A_\varphi(\lambda)$, to seconds of arc, for each integer degree of λ; for $\varphi = 0°$ (this table is ascribed to Thābit bin Qurra (**93**), and takes $\epsilon = 23;33°$), and for $\varphi = 33;25°$, the latitude of Baghdād.

$\bar{h}(\lambda_s)$ computed to seconds of arc, for each integer degree of the argument and for $\varphi = 36;0°$.

The same *Jadwal at-Taqwīm*, but with columns transposed, as is described in §8, C below, is found also in this version. In addition, two more elaborate sets of ten functions each, of the same general character, are displayed here. These are carried to two or three fractional places for the range of argument 1°, 2°, 3°, $\cdots$, 90°. Such of the ten as are standard trigonometric functions are noted above.

D. EQUATION OF TIME

There is no table of $E(\lambda_s)$.

E. MEAN MOTIONS

The mean motion tables are set up as with the Istanbul version, §8 below, except that the base year is 511 A.H. running to 841 or 691, and tables are sometimes carried to six or eight fractional places. The length of the tropical year is given as 6, 5; 14, 27, 16, 36, 28, 18 days.

Of the two following tables of apogees the first (f. 17) is specifically ascribed to the Mumtaḥan Zīj (**51**), but is for 876 A.H.

☉	92; 24, 3, 3, 5, 1, 4, 2, 17, 5, 2°
♄	252; 34, 23, 8, 6, 11, 40, 1, 4, 0, 1
♃	182; 18, 23, 7, 30, 18, 1, 0, 5, 8, 6
♂	134; 14, 23, 5, 3, 9, 3, 8, 7, 20, 7
♀	92; 24, 3, 1, 0, 20, 8, 6, 9, 40, 1
☿	213; 44, 3, 12, 3, 50, 0, 4, 12, 50, 3

The second (f. 28) gives no indication as to the date for which it is computed or for its origin. It is

☉	79; 30, 23, 2, 43, 53°
♄	239; 41, 23, 2, 43, 53
♃	169; 23, 23, 2, 43, 53
♂	121; 24, 23, 2, 43, 53
♀	same as ☉
☿	197; 51, 23, 2, 43, 53

Presumably the common fractional endings above are due to the addition of a common precessional constant to some other set of apogees. For instance, neglecting seconds and beyond, the difference between corresponding entries in the two tables above runs about 12; 53°. This is precessional motion for about 876 lunar years. Thus the epoch of the second table can be conjectured to be 1 A. H.

A table of apsidal motion (all apogees moving together) for each Hijra year, months, and days, to two and three fractional places, has the approximate rate of 0; 0, 0, 8, 58° per day.

E. PLANETARY EQUATIONS

These tables also are like the corresponding ones of the Istanbul version, except that the solar equation is carried to thirds of arc.

G. PLANETARY LATITUDES

The arrangement is the standard Ptolemaic one, except that β_m is 4; 46° as in §8, G below. This, according to al-Khāzinī (in **27**) is the value determined by the Mumtaḥan (**51**) observations.

H. STATIONS AND RETROGRADATIONS

There is a table of stations, computed to minutes for each six degrees of the argument.

I. PLANETARY SECTORS

None.

J. PARALLAX

This zīj reproduces the *Almagest* tables of solar and lunar parallax in the altitude circle.

There is a table (f. 153) called "Lunar Parallax for Visibility of the Crescent" consisting of three functions, crudely computed as though by linear interpolation in segments of irregular length. They are:

(*a*) Some sort of interpolation scheme for variation due to lunar anomaly.

(*b*) 0;6 sin θ.

(*c*) 23;35 sin θ (hence an approximate declination table).

The method of application of these tables is unknown to the present writer, and they may properly belong in *L* below.

K. ECLIPSE THEORY

The eclipse tables are the same as those in §8, K below.

L. VISIBILITY PROBLEMS

For the latitude of Baghdād and for each of the seven climates there are planetary visibility tables of the same form as those in the *Almagest*, but with entries differing from them and also from the tables described in §12, L below. (*Cf. Nallino, Batt.*, vol. II, pp. 255–269.)

M. GEOGRAPHICAL TABLES

None.

N. STAR TABLES

There is a table of thirty stars giving for each (λ, β) and (α, δ) for the year 304 (presumably A. H.). The latitudes are the same as those for corresponding stars of §5, N, whence it is to be inferred that these coordinates have been computed from the results of the Mumtaḥan observations. Also given are meridian altitude, the degree of the simultaneously rising point of the ecliptic, with its ascension, half the daylight arc of the star, and half its "diameter" (*quṭr*, i.e., the radius of the small circle traced out by it in the course of the daily rotation, Cos δ. This is the "day-radius" of Hindu astronomy.)

O. ASTROLOGICAL TABLES

There is a table (f. 157) for computing the instant of *nativity transfer* based on the Hijra calendar, for 1, 2, 3, ···, 30, 60, 90 years and individual months. All entries are to three fractional places. The *excess of revolution* on which the table is based is 86; 43, 39, 37°, to which corresponds a tropical year of 6, 5; 14, 27, 16, 36, 10^d. Along the margin of the same folio are three more such parameters given, which with the associated tropical years are shown below:

86; 41, 25, 14, 9°	6, 5; 14, 26, 54, 12, 21, 30^d
86; 43, 39, 36, 47	6, 5; 14, 27, 16, 36, 7, 50
86; 35, 13, 46	6, 5; 14, 25, 52, 17, 40

The first is said to be the result of the observation at Damascus; the second that of the Shamāsī observations in Baghdād; and the third the value of Yaḥyā ibn Abī Manṣūr. Doubtless all are to be considered Mumtaḥan observations. Evidently the parameter used in the table above, as well as that of the table mentioned in §5, O, is a rounded-off number obtained from the Shamāsī parameter. Yaḥyā's value rounds off to the very common 6, 5; 14, 26. It is also close to the 6, 5; 14, 25, 52, 20 implied by *Bīrūnī* (*Chron.*, transl., p. 141), and to the 6, 5; 14, 25, 52 attributed to Nairīzī by Ibn Yūnis (*Caussin*, p. 74).

P. MISCELLANEOUS

There is a table (f. 150) of the apparent diameters of the seven planets, to two places, for each six degrees of the anomaly, apparently from Hindu sources. (*Cf. Sūrya-Siddhānta*, p. 196.)

There is an interpolation table for lunar and solar distances, computed to two places, for each two degrees of the anomaly.

8. ABSTRACT OF 16, THE ISTANBUL COPY OF A ZĪJ OF *ḤABASH AL-ḤĀSIB*, *c.* 850

Unlike **15** (§7) above, the general impression given by this manuscript is that it is much more homogeneous than the other purported copies of early zījes. Exceptions to this are noted in the sequel.

A. CHRONOLOGY

Text and tables are present for the Hijra, Yazdigird, Seleucid, and Coptic calendars. The era of Nabonassar is mentioned. In addition there is a table of the lunar mansions, and a royal canon of the caliphs, down to al-Muṭī'lillāh al-Faḍl ibn al-Muqtadir (*c.* A.D. 950).

B. TRIGONOMETRIC FUNCTIONS

There are tables of:

Sin θ, to four places, for $\theta = 0;0°, 0;15°, 0;30°, \cdots, 90;0°$,

Tan θ, to two fractional places, for $\theta = 0;0°, 0;30°, 1;0°, \cdots, 89;0°$,

Tan θ, to two fractional places, for $\theta = 0°, 1°, 2°, \cdots, 90°$,

and 24 tan θ to two fractional places, for $\theta = 0°, 1°, 2°, \cdots, 90°$.

C. SPHERICAL ASTRONOMICAL FUNCTIONS

There are also tables of:

$\delta(\theta)$, to seconds of arc, for each integer degree of θ, with $\epsilon = 23;35°$.

$A_\varphi(\lambda)$, for each integer degree of λ, for $\varphi = 0°$ and $\varphi \approx 34°$ (neither the location nor the precise latitude is specified), also for each of the seven climates to minutes of arc.

There are tables giving, for each integer degree of the argument, $\bar{h}(\lambda_s)$ to seconds of arc for $\varphi \approx 34°$, and to minutes of arc for each of the seven climates.

There are two blocks of tables each called *Jadwal at-Taqwīm* (Table (for) the True Position). They consist of four functions each, to be used in spherical astronomical computations, computed to two fractional places for arguments of $1°, 2°, 3°, \cdots, 90°$. These two sets contain essentially the same functions, but with slightly different values. Probably they are the work of two computers using slightly different basic parameters.

A short work of Abū Naṣr Manṣūr (**77**) (*Bankipore*, vol. XII, p. 66, Ms. 2468, 8) is entitled "On the Proofs of the Operations of the 'Table (for) the True Position' in the Zīj of Ḥabash al-Ḥāsib." From this it is to be concluded that the table in question was part of the Ḥabash Zīj in its original form. The manuscript has been published as part of *Risā'il Abī Naṣr Manṣūr* . . . , Hyderabad-Deccan, 1948.

D. EQUATION OF TIME

There is no table of $E(\lambda_s)$.

E. MEAN MOTIONS

These tables are, in general, computed to two fractional places, occasionally to three. Positions of $\bar{\lambda}_s$, $\bar{\lambda}_m$, a_m, λ_n, and $\bar{\lambda}$ of the superior planets, and a of the inferior are given for years 1, 31, 61, $\cdots$, 691 A.H. in the original hand (in some cases this is extended to 991 A.H. by a later user). The motions of these same means in 1, 2, 3, $\cdots$, 30 Hijra years; 1, 2, 3, $\cdots$, 29 days; 1, 2, 3, $\cdots$, 24 hours; and 10, 20, 30, $\cdots$, 60 minutes are also given.

Precessional motion is given for each Hijra year, the (approximate) rate being 0; 0, 54, 44, 20° per Hijra year.

F. PLANETARY EQUATIONS

The arrangement is standard. All tables are computed for each degree of the arguments, to seconds of arc in the case of e_s, to minutes for all others.

G. PLANETARY LATITUDES

The tables of planetary latitudes are for each 6° of the arguments and carried to minutes of arc. However

$$\beta_m = 4;46° \sin(\lambda_m - \lambda_n),$$

the table of which is computed for each degree of the argument, to seconds of arc.

H. STATIONS AND RETROGRADATIONS

As in §7, H.

I. PLANETARY SECTORS

None.

J. PARALLAX

A table called simply "Parallax" gives 24 sin θ to two places. This, disregarding the sexagesimal point, yields values approximately half the corresponding entries in the Khwārizmī P_θ table (§6, J). The accompanying text describes a method for computing parallax components, including a very neat recursion algorism for producing the Khwārizmī P_λ. From its resemblance to the methods of the *Sūrya-Siddhānta* it is taken to be of Hindu origin.

This zīj also contains the *Almagest* table of solar and lunar parallax in the altitude circle.

K. ECLIPSE TABLES

There is a table of $\lambda_s'(\bar{\lambda}_s)$ and $\lambda_m'(a_m)$ in degrees per hour for each degree of the argument; also in degrees per 0; 1 days for each four degrees of the arguments, all to thirds of arc.

There is a table of the motion of $\bar{\lambda}_s$, λ_n, and the mean elongation, to seconds of arc, for 0;1, 0;2, 0;3, ···, 1;0 days.

There are tables of mean conjunctions and oppositions. The argument runs through Hijra years 1, 31, 61, ···, 871, and changes in the functions are given for 1, 2, 3, ···, 30 Hijra years and for each lunar month. Functions tabulated are:

λ of the mean conjunction or opposition, to three fractional places of degrees,
t, to three fractional places of days,
a, to minutes of arc, and
λ_n to minutes of arc.

A table of lunar eclipses is said, in the zīj, to utilize the method of Ptolemy. For eclipse magnitudes of 0, 1, 2, ···, 21 digits and for the moon at greatest and least distances the table gives β_m and times of immersion and (where applicable) totality, to two places. An interpolation scheme is set up for intervals of six degrees of a_m.

A table of solar eclipses is arranged as above, the argument proceeding to 12 digits only, and there is no column for time of totality.

There is a table of eclipse inclination, of immersion and totality, tabulated for each digit to twenty-one, for both lunar and solar eclipses.

In addition to these tables and the accompanying explanatory text, there is a long explanation of an alternative method of eclipse computation employing graphical and numerical methods largely without the use of tables. These sections are probably of Hindu origin and will be well worth detailed study.

L. VISIBILITY TABLES

The *Almagest* planetary visibility table is reproduced in this zīj.

There is in addition a set of tables giving, for each climate, each planet, and each zodiacal sign a value of the arc of visibility computed to minutes of arc. The entries are different from those of §12, L below.

M. GEOGRAPHICAL POSITIONS

None.

N. STAR TABLES

For twenty-four stars the coordinates (λ, β) and (α, δ) are given, to minutes of arc, these being the Mumtaḥan (51) observations of 214 A.H.

O. ASTROLOGICAL TABLES

This zīj has two tables, differing from each other and from that of Khwārizmī (§6, O which see), giving, to seconds of arc, the position of the mean sun when the true sun enters each of the signs, and various decans.

9. ABSTRACT OF 55, *AZ-ZĪJ AṢ-ṢĀBI'* OF *AL-BATTĀNĪ*, *c.* 900

(Page references are to *Nallino, Batt.*, vol. II)

A. CHRONOLOGY

Full material (pp. 7–17) is given for only two calendars, that of the Hijra, and the Seleucid. Use is made of the Coptic (i.e., Egyptian) year, but only for planetary computations. Among the spurious tables at the end of the zīj are tables (pp. 300–301) said to have been taken from the zīj of Maslama, presumably al-Majrīṭī, the editor of **21**. The tables are of standard type, of initial week days of months and years of the Hijra and Yazdigird calendars. Since tables closely resembling these appear in **21** (p. 110), we do not regard this as sufficient evidence for crediting Maslama with an independent zīj.

An extensive royal canon (pp. 1–6) commences with Nabonassar and runs through the Achaemenian, Ptolemaic, Roman, and Byzantine dynasties. In this the durations of reigns are given in years. This is followed by a canon of the Caliphs in years and days down to al-Muktafī b'illāh (294 A.H.). The canon (in the manuscript used by Nallino) was extended by someone other than Battānī to include al-Muṭī' li'llāh (*c.* A.D. 950).

B. TRIGONOMETRIC TABLES

There are tables of:

Sin θ, to three places, for $\theta = 0; 0°, 0; 30°, 1; 0°, \cdots, 180; 0°$, (pp. 55–56) and 12 cot θ, to one fractional place, for $\theta = 1°, 2°, \cdots, 90°$ (p. 60).

C. SPHERICAL ASTRONOMICAL FUNCTIONS

There is also a table (pp. 57–58) of $\delta(\theta)$ to seconds of arc, for each integer degree of θ, with $\epsilon = 23; 35°$.

The following tables of ascensions (pp. 61–71) are displayed, all computed to one fractional place:

$A_0(\lambda) + 90°$, for each integer degree of λ,

$A_{36;0°}(\lambda)$ for each integer degree of λ, this being for the latitude of Raqqa, the location of al-Battānī's observatory.

$A_\varphi(\lambda)$ for $\lambda = 0°, 10°, 20°, \cdots, 360°$ and φ's such that in those locations the maximum length of daylight is 13;0, 13;15, 13;30, ···, 16;0 hours, also for $\varphi = 21;40°$ (Mecca), 33;9° (Baghdād), and 36;40° (Ḥarrān).

For the latitude of Mecca, Baghdād, Ḥarrān, and Raqqa there are associated with the corresponding tables of ascension, entries giving $\bar{h}(\lambda_s)$ for the same domain of the argument and with the same precision as the corresponding ascensions.

There is also a table (p. 59) showing (max $D - 12$) as a function of latitude, computed to minutes of arc for $\varphi = 0;30°, 1;0°, 1;30°, \cdots, 60;0°$.

D. EQUATION OF TIME

There is a table of $E(\lambda_s)$, to minutes of arc for each integer degree of the argument.

E. MEAN MOTIONS

Solar and lunar mean motions are all carried to two fractional places, those of the planets to one. The locations of $\bar{\lambda}_s$, λ_m, a_m, and λ_n are shown for years 931, 951, 971, ···, 1631 of the Seleucid era (for the planets the table is carried to 1591 only). Totals of these means in 1, 2, 3, ···, 20 (Seleucid) years; 1, 2, 3, ···, 12 months; 1, 2, 3, ···, 31 days; and 1, 2, 3, ···, 24 hours are also shown (pp. 73–77, 102–107).

Complete tables of mean motions are also given for the Hijra calendar (pp. 19–28) in intervals of 30 lunar years, the range of the arguments otherwise corresponding to the arrangement for the Seleucid calendar. See §10, E below.

F. PLANETARY EQUATIONS

Two equations, those of the sun and of Venus, differ from the values of the *Almagest*, otherwise the tables (pp. 78–83, 109–137) are like those of, say, §8, F.

G. PLANETARY LATITUDES

These are completely Ptolemaic, with max $\beta_m = 5°$ as in the *Almagest* (pp. 140–141).

H. STATIONS AND RETROGRADATIONS

As in §7, H (pp. 138–139).

I. PLANETARY SECTORS

None.

J. PARALLAX

Battānī gives the *Almagest* table (pp. 93–94) of solar and lunar parallax in the altitude circle, but with emendation of some values.

He also has the Theon tables (pp. 95–101) of parallax components, for all seven climates.

K. ECLIPSE THEORY

There is a table (p. 88) of $\lambda_s'(\bar{\lambda}_s)$ and $\bar{\lambda}_m'(a_m)$ to two places for each six degrees of the arguments.

There are two tables (pp. 29–32, 84–87) of mean conjunctions and oppositions, one for arguments of cycles of twenty-five Egyptian (Coptic) years from 915 to 1690 and for motions in years and months within the cycle. The other set is for cycles of twenty-four Seleucid years, ranging from 376 to 1623. In both cases functions tabulated are t, λ, a_m, and λ_n, of the conjunction or opposition, all to seconds of arc or time.

The Ptolemaic digit conversion table (p. 89) is reproduced. There is also a table giving the eclipse inclination as a function of magnitude.

There are tables (p. 90) of lunar eclipses, one for the moon at greatest distance, one for least. The argument is each integer digit of eclipse magnitude up to 21;31,30 digits in the first case, to 21;36 digits in the second. Functions tabulated are: β_m, duration of immersion, and duration of totality, all to two places. There is an interpolation scheme for intermediate lunar distances.

There are two analogous tables (p. 91) for solar eclipses, laid out as above except that the limits of magnitude are 11;23,30 digits and 12;33 digits respectively, and there is no column for duration of totality.

L. VISIBILITY THEORY

There is a planetary visibility table (pp. 142–143) set up in the same manner as that of the *Almagest*, but with different entries. There is only explanatory material (vol. I, pp. 266–272) on first visibility of the lunar cresent. In the Sanjarī Zīj (27, see §12, L below), however, appears a table of crescent visibility computed according to al-Battānī's conditions.

M. GEOGRAPHICAL POSITIONS

There is a table (pp. 33–54) giving the latitude and longitude of 273 locations, to minutes of arc.

N. STAR TABLES

One table (pp. 144–177) is for epoch 1191 Alexander (i.e., of the Seleucid era, A.D. 888) and gives for 533 stars latitudes, longitudes, and magnitudes. In general the latitudes are those of the *Almagest*, the longitudes corrected for precession.

Another table (pp. 178–186) has entries for seventy-five stars, arranged in order of magnitude, for epoch 1211 Alexander (A.D. 900), and including the declination, the meridian altitude, half the daylight arc, the degrees of culmination, rising, and setting, all to minutes of arc for $\varphi = 36°$ (Raqqa).

O. ASTROLOGICAL MATERIAL

Among the spurious tables at the end of the zīj is an "astrological rose" (p. 299), a circular diagram showing the rulers of the houses, triplicities, exaltations, etc.

There is a table (p. 188) of the *excess of revolution* for 1, 2, 3, ⋯, 20, 40, 60, ⋯, 200 years. The parameter is 86; 36° per year, corresponding to a tropical year of 6, 5; 14, 26 days (*cf.* §5, O).

P. MISCELLANEOUS

There are tables (p. 188) giving the altitude and azimuth of the sun at each of the twelve unequal hours, the geographical location being at the latitude of Raqqa, and the sun being assumed in the first of Capricorn and then in the first of Cancer.

10. ABSTRACT OF 9, *AZ-ZĪJ AL-JĀMI'*, OF *KŪSHYĀR*, *c.* 1000

A. CHRONOLOGY

This work has text and tables for the Hijra, Yazdigird, and Seleucid calendars. There is also a table of the lunar mansions, giving latitude and longitude to minutes of arc.

B. TRIGONOMETRIC FUNCTIONS

There are tables of:

Sin θ, to three places, $\theta = 0;0°, 0;1°, 0;2°, \cdots, 90;0°$,

Tan θ, to three places, $\theta = 0°, 1°, 2°, \cdots, 60°$, with tabular differences.

Tan θ, to three places, $\theta = 0°, 1°, 2°, \cdots, 45°$, with tabular differences.

$\left.\begin{array}{r} 7\cot\theta, \\ 12\tan\theta, \\ 12\cot\theta, \\ \text{Vers}\,\theta, \end{array}\right\}$ to three places, for each integer degree of the argument, with tabular differences.

C. SPHERICAL ASTRONOMICAL FUNCTIONS

There are also tables of:

$\delta_1(\theta)$ and $\delta_2(\theta)$, to two fractional places, for each integer degree of θ, with tabular differences, $\epsilon = 23;35°$, and Tan $\delta_1(\theta)$, to two fractional places, for each integer degree of θ.

Tables of ascensions, all computed to seconds of arc, and for each integer degree of λ are:

$$A_\varphi(\lambda) \quad \text{for} \quad \varphi = 0° \quad \text{and} \quad \varphi = 35;30°,$$

and following the colophon, hence not part of the original zīj, are:

$$A_0(\lambda) + 90°,$$

and $A_\varphi(\lambda)$ for $\varphi = 30;5°$ (Bardishīr) and $\varphi = 29;30°$.

There is a table, computed to minutes of arc, of

$$\Delta D(\lambda_s), \text{ for each integer degree of } \lambda_s.$$

After the colophon of the zīj proper are tables of $\Delta D(\lambda_s)$ and $\bar{h}(\lambda_s)$, set up as above, for $\varphi = 36;15°$.

There is a table, computed to minutes of arc for each integer degree of the argument, of

$$\max h_s(\lambda_s), \quad \text{for} \quad \varphi = 29;30°.$$

D. EQUATION OF TIME

There is a table of $E(\lambda_s)$, to seconds of time, for each integer degree of the argument.

Following the colophon are two more such tables, for each six degrees of the argument, which is $\bar{\lambda}_s$ for the one and $\bar{\lambda}_m$ for the other.

E. MEAN MOTIONS

This zīj has a table of the positions of all the means at the epoch of the Yazdigird calendar, the mean motions in twenty Persian (i.e. Egyptian) years and for a single day, all the above carried to six fractional places. They are stated to be from al-Battānī, and they are in fact the same as these given (implicitly) in **55**.

These base parameters were used to compute mean motion tables of the usual type, but in the Yazdigird rather than the Hijra or Seleucid calendars. These are to two or three fractional places. Mean positions are given for Yazdigird years 1, 21, 41, ⋯, 481, and motions of these means in 1, 2, 3, ⋯, 20, 40, 60, ⋯, 100, 200, 300, ⋯, 600 years; 1, 2, 3, ⋯, 12 months; 1, 2, 3, ⋯, 24 days. There is also a table for correcting mean positions according to the geographical longitude of the observer.

The planetary apogees are those of Battānī (**55**) but with positions as of the beginning of the Yazdigird era.

After the colophon of the zīj proper are a few additional folios of tables among which is a listing of all the mean motions in degrees per day to six fractional places. These are ascribed to az-Zīj al-Fākhir (**44**) and are the same as those noted above. This confirms the statement of al-Fārisī (**54**, *Lee*) that the elements of both zījes are taken from al-Battānī (**55**). Since these parameters are not explicitly reported in *Nallino, Batt.*, we reproduce them below:

☉		0; 59, 8, 20, 46, 56, 14° per day
☾	(mean)	13; 10, 35, 2, 7, 17, 10
	(anomaly)	13; 3, 53, 56, 17, 51, 59
	(The above value is that of Ptolemy in the *Almagest* [transl. of Manitius, vol. I, p. 210].)	
	(nodes)	0; 3, 10, 37, 18, 40, 26
♄		0; 2, 0, 35, 51, 48, 3
♃		0; 4, 59, 16, 55, 54, 57
♂		0; 31, 26, 40, 15, 11, 13
♀	(anomaly)	0; 36, 59, 29, 28, 42, 45
☿	(anomaly)	3; 6, [2]4, 7, 45, 53, 33

There is a table of apsidal motion, to seconds of arc, for the same domain as the arguments of the mean motion.

After the colophon, hence probably from az-Zīj al-Fākhir (**44**) is another table of apogees and of apsidal motion. Both sets of tables are said to be obtained from those of al-Battānī, and this is in fact the case.

F. PLANETARY EQUATIONS

These tables are as in §9, F above except that the solar and lunar equation tables have tabular differences.

G. PLANETARY LATITUDES

Same as §9, G.

H. STATIONS AND RETROGRADATIONS

As in §7, H.

I. PLANETARY SECTORS

None.

J. PARALLAX

There is a table of solar parallax in the altitude circle, computed to two places at intervals of three degrees of the argument. Maximum parallax is 0; 3, 0°.

One of the Theon parallax components tables is given, that for the third climate.

K. ECLIPSE THEORY

In the zīj proper there is a table of $\lambda'(\bar{\lambda}_s)$ and $\lambda_m'(a_m)$ in degrees per hour, $r_s(\bar{\lambda}_s)$, $r_m(a_m)$, and $r_u(a_m)$ for each twelve degrees of the arguments, all to seconds of arc, at syzygies.

There is a table of earth–moon distance, to minutes, taking the lunar deferent radius as 1, 0, for each six degrees of a_m and 0°, 5°, 10°, · · ·, 35° double-elongation.

The Ptolemaic digital conversion table is reproduced.

After the colophon, hence presumably not the work of Kūshyār, are:

A lunar eclipse table ascribed to Bīrūnī (**59**) in which there are two arguments:

(1) Distances between the longitudes of conjunction and node of 0; 0°, 0; 30°, 1; 0°, · · ·, 11; 30°, and
(2) Ranges of lunar daily motion of from 12; 0° to 12; 20°, from 12; 21° to 12; 48°, from 12; 49° to 13; 11°, from 13; 12° to 13; 45°, from 13; 46° to 14; 16°, and from 14; 17° to 14; 48°.

Functions tabulated are times of immersion and totality, to minutes.

Bīrūnī's solar eclipse table with arguments

$$\beta_m = 0', 1', 2', \cdots, 34', \quad \text{and}$$
$$\lambda_m' = 12; 0°, 12; 24°, 12; 48° \text{ per day},$$

yielding immersion time and magnitude, to minutes.

L. VISIBILITY CONDITIONS

Following the colophon is a table, apparently of coefficients for modifying some sort of crescent visibility conditions. The arguments are each six degrees of a_m and two or three values of the double elongation, the latter garbled. Entries are to minutes.

No other visibility tables appear.

M. GEOGRAPHICAL LOCATIONS

There is a table layout, but left blank, for forty-five localities, longitude to be reckoned from the Fortunate Isles, coordinates to minutes of arc.

After the colophon appears a table of the latitude and longitude of ninety-one localities, to fives of minutes, longitudes reckoned from the Fortunate Isles.

N. STAR TABLES

A star table was laid out in this copy, but no star names or coordinates were ever entered. There are places for thirty stars, for latitude and longitude as of the year 301 Yazdigird, for their astrological temperaments, and magnitudes.

Following the colophon is a table of thirty stars giving (β, λ) (for year 1293 of Alexander) and (α, δ), most coordinates terminating in multiples of five minutes. Astrological temperaments also are given. The latitudes are the same as those of the *Almagest*, whence it can safely be concluded that all the other coordinates have been obtained from the Ptolemaic values with suitable allowance for precession.

O. ASTROLOGICAL TABLES

There is a table layout headed *at-Tasyīrāt*, but this was never filled in in this copy.

Following the colophon is:

A table for the equalization of the houses, to three places for each degree of the ascendant.

Two tables of nativity lots giving, for each of the planets, and for diurnal and nocturnal nativities, the associated sign, the Egyptian terms (to five places), and the lords of the triplicities and decans.

11. PARTIAL ABSTRACT OF **59**, *AL-QĀNŪN AL-MAS'ŪDĪ*, OF *AL-BĪRŪNĪ*, 1030

Both the *Brit. Mus., Suppl.* and the *Berlin* catalogues give (in Arabic) full tables of contents of this work. Because of its unusual importance we will abandon, for this section only, the master outline given in §4 and will follow the order of topics as presented in the actual zīj. Of the numerous extant copies of the Canon, a microfilm of only one, *Bodl. II, 2*, Ms. 370 (Bodl. 516), was in the possession of the present author at the time of writing. Although this is an old and valuable copy, it contains only the first six out of a total of eleven treatises in the whole work. Notices of the tables in the last five treatises will, therefore, be incomplete.

Treatises I and II include a discussion of fundamental astronomical concepts and the chronological material contained in the zīj. Full tables are given for the Hijra, Seleucid, Yazdigird, Jewish, Coptic, Julian, and Magian (i.e. Zoroastrian) calendars, together with tables of holidays of the various sects, and an extensive royal canon. This material goes over much of the ground covered in Bīrūnī's independent "Chronology of Ancient Nations."

Treatise III is the trigonometric section of the Canon. As remarked above, this treatise has been ably translated (or rather paraphrased) into German by *Schoy* (*Mas'ūdī*). There are two tables, one giving Sin θ to four places for each 15′. There are two columns of tabular differences associated with each value of the argument, one giving the increment for a change in the argument of one minute, the other the increment for a quarter-degree change. The second table is of Tan θ, computed to five places, and with columns of first and second differences.

Treatise IV is given over to topics in spherical astronomy. All the tables in this treatise have entries for each integer degree of the arguments and are computed to four-place accuracy.

The first chapter defines the constant ϵ, Bīrūnī taking for it the value 23; 35°. There is a publication by Farooq, M., "Al-Kanun-Ul-Masudi," The Muslim University Press, Aligarh, India, 1929, which contains the Arabic original, translation, and commentary on a small part of the Canon. This may be the first chapter of Treatise IV. The second chapter (translated in *Schoy, Mas'ūdī*) is devoted to declinations of points on the ecliptic, and is followed by tables of δ_1, and δ_2. The third chapter deals with right ascensions and is followed by a table of $A_0(\lambda)$.

Chapters 4, 5, and 6 give methods of transforming from equatorial to ecliptic coordinates and vice versa.

Chapters 7–17 inclusive are concerned with gnomonics. Of these, Chapters 8, 11, 14, and 17 were translated by Schoy, the last three in *Mas'ūdī*, the first in *Annalen der Hydrographie und Maritimen Meteorologie*, vol. 53 (1925), pp. 41–47. Chapter 10 is followed by a table giving, for the latitude of Ghazna (φ = 33; 35°), the duration of daylight in equal hours, the length of the unequal hours, and the meridian altitude of the sun. Chapter 11 has a table of Tan max h_s and 12 Tan max h_s.

Chapter 18 has a table of $A_\varphi(\lambda)$ for Ghazna.

The rest of the twenty-six chapters of the treatise give astronomical methods of time determination, the fixing of the astrological centers, transformations of times and ascensions for changes of geographical location, and the determination of the ascendant at the *cupola of the earth* (Ujain).

Treatise V is taken up with geodetic problems. The first chapter describes the determination of the longitudes of a locality by simultaneous observations of eclipses. The second chapter, translated by *Schoy* (*Geogr.*) solves the same problem by use of the distance from a point of known coordinates. Chapter 3 describes how to determine the terrestrial distance between two points of known coordinates. Chapter 4 gives a method of determining the coordinates of a locality in terms of its distances from two fixed localities. The remaining seven chapters are missing in the Bodleian copy. Chapters 5 and 6, the latter of which is translated in *Schoy, Mas'ūdī*, dispose of the determination of the azimuth of one locality with respect to another. When one of the localities is Mecca the azimuth is that of the *qibla*, the Muslim direction of prayer. Excerpts and translations from Chapter 7 (as well as a passage from Chapter 2, Treatise I) dealing with the circumference of the earth, are to be found in "Muslim Researches in Geodesy" by S. H. Barani, pp. 1–52, of the "Al-Bīrūnī Commemoration Volume," Iran Society, Calcutta, 1951. The whole of the chapter had previously been translated in *Schoy, Geogr.* Chapters 8 and 9 deal with parallels of latitude and the location of the climates. A one-folio table follows the latter. Chapter 10 consists mainly of a six-folio table of geographical locations.

Treatise VI deals primarily with the problem of solar motion. The first chapter, however, discusses the determination of time differences between localities of different longitude. The second chapter, translated in *Schoy, Geogr.* is a computation of the difference in longitude between Alexandria and Ghazna. (This chapter has been retranslated by J. H. Kramers in pp. 177–193 of the "Al-Bīrūnī Commemoration Volume" mentioned above.) The solar mean motion is next discussed. Included in the sixth chapter is a table in which twenty-three observations of equinoxes are recorded beginning with those of Hipparchus and Ptolemy and ending with two by Bīrūnī himself. Publication of this table with the accompanying text would enable the computation of the solar mean motion as determined by many early Islamic observers. Most of these parameters are already known from other sources, and the known ones could be used to control the computation of those unknown. Chapters 7 and 8 deal with the motion of the solar apogee. Bīrūnī settles on a value of 0; 0, 0, 8, 34, 31, 25, 1° per day for the latter. He gives the mean motion of the sun as 0; 59, 8, 12, 7, 56, 33° per day. Chapter 9 has a table of the solar mean motion and apsidal motion carried to six fractional places for 1, 2, 3, ···, 30 Persian (i.e., Egyptian) years, and for the same range of Persian months and days. Positions of the mean sun and apogee are given for Yazdigird years 400, 430, 460, ···, 820. Chapter 11 has a table of the solar equation (maximum is 1; 59, 31, 40, 30°) to four places for each degree of the argument, with tabular differences. The treatise concludes with a chapter on the equation of time.

Treatise VII is devoted to the lunar motion and has the usual tables of mean motion, equations, and latitude. The concluding chapters, 10 and 11, however, have text but no tables on the determination of latitude and longitude components of the lunar and solar parallax respectively. The last chapter also has a section each on the apparent lunar and shadow diameters, and on the distance of the sun from the earth.

Treatise VIII presents eclipse and lunar visibility theory in a total of seventeen chapters. There are chapters on: lunar and solar angular velocities and rates of elongation (with one folio of tables), mean conjunctions and oppositions, the lunar shadow (with one folio of tables, *cf.* §10, K below), eclipse limits, color of lunar and solar eclipses, duration of eclipses, eclipse magnitudes, and eclipses occurring near sunrise or sunset. Chapters 12 through 17 have as topics: the illumination (of the lunar disk), definition of twilight and dawn, visibility of the crescent, the lunar mansions, and lunar days. The latter are the "tithis" of Hindu astronomy (*cf. Neugebauer*, p. 123).

Treatise IX concerns the fixed stars. In its nine chapters are included as topics: the differences between the planets and the fixed stars, the arrangement of the latter according to the inhabitants of various regions of the earth, motion of the fixed stars, a nineteen-folio star table, risings and settings of the fixed stars, and stars associated with the lunar mansions by the Arabs and the Hindus.

Treatise X has to do with the motions of the five planets, and comprises thirteen chapters. Except for the material on visibility, the tone is strongly Ptolemaic, and the topics and tables are the standard ones.

Treatise XI is purely astrological, its twelve chapters including two methods of equalization of the houses (the well-known one and Bīrūnī's own), aspects, projections of the rays (again by the standard method and by the one Bīrūnī developed), the apheses and progressions (with two-folio and six-folio tables respectively), year-transfers and nativity-transfers, and deferent and epicycle sectors.

12. ABSTRACT OF 27, *AZ-ZĪJ AS-SANJARĪ*, OF *AL-KHĀZINĪ*, *c.* 1120

A. CHRONOLOGY

The chronological section of this zīj is very extensive. It has material on the Hijra, Yazdigird, Seleucid, Jewish, Soghdian, and Hindu calendars. There are tables of Muslim, Zoroastrian, Christian, and Jewish fasts and holidays as well as regnal tables of the Babylonian, Achaemenian, Macedonian, Coptic (Roman Emperors), Sasanian, Umayyad, Abbasid, Byzantine, North African, Buyid, and Seljuk dynasties. There is a chronological table of prophets, and a table of the lunar mansions.

B. TRIGONOMETRIC FUNCTIONS

There are tables, all computed for each integer degree of the argument, for

Sin θ, with first and second differences,
Vers θ, to 180°, with first differences,
Cot θ, with first differences, } to three places,

and $6\frac{1}{2}$ cot θ,
7 cot θ,
12 cot θ, } to two places.

C. SPHERICAL ASTRONOMICAL FUNCTIONS

There are also tables of:

$\delta_1(\theta)$ and $\delta_2(\theta)$ to seconds of arc for each integer degree of θ, with $\epsilon = 23;35°$.

Tables of ascensions, all computed to minutes of arc for each integer degree of λ, are:

$A_\varphi(\lambda) + 90°$, with tabular differences, for $\varphi = 0°$, and $\varphi = 37;40°$ (for Marv, the location of al-Khāzinī's observatory), and $A_\varphi(\lambda)$, for all seven climates, and for $\varphi = 33;25°$ (Baghdād), $\varphi = 90° - \epsilon = 66;25°$, $\varphi = 76;4°$, $\varphi = 37;40°$, the last with tabular differences.

There are tables, computed to minutes, of $\bar{h}(\lambda_s)$ and $\Delta D(\lambda_s)$ for each integer degree of λ_s, $\varphi = 37;40°$ (Marv), and (max D − 12) and sin (max D − 12) for $\varphi = 1°, 2°, 3°, \cdots, 60°$.

D. EQUATION OF TIME

There is a table of $E(\lambda_s)$, to seconds of time, for each integer degree of the argument.

E. MEAN MOTIONS

In keeping with the other parts of this magnificent work, the section on mean motions is unusually complete and precise. All the basic mean motions, i.e. both mean longitude and anomaly for all planets, the motion of the lunar nodes, and the rate of double elongation for the moon are given to seven or more fractional places, in degrees per day, and revolutions per day. The former are reproduced below:

☉	0; 59, 8, 20, 33, 53, 4, 29, 40°
☾	13; 10, 35, 2, 0, 41, 28, 38, 50
♄	0; 2, 0, 36, 4, 43, 2, 8
♃	0; 4, 59, 16, 19, 53, 47, 11, 20
♂	0; 31, 26, 39, 36, 34, 5, 16, 50
☊	0; 3, 10, 37, 38, 17, 2, 57, 30
☾ (anomaly)	13; 3, 53, 56, 12, 33, 51, 26, 30
♀ (anomaly)	0; 36, 59, 28, 43, 1, 37, 38, 20
☿ (anomaly)	3; 6, 24, 7, 9, 39, 35, 45, 50

There are tables showing the amounts of all these motions in degrees carried to eight fractional places for 1, 2, 3, ···, 60 days.

Al-Khāzinī makes precessional and apsidal motion the same and puts it at 0; 0, 0, 8, 57, 38, 45, 42, 30° per day, which comes (approximately) to 0; 0, 52, 56° per Hijra year.

The initial points of all the mean motions, to seconds of arc, for longitude 90° from the "shore of the western sea" (probably from the Fortunate Isles) are given for the epoch of the Hijra, Yazdigird, and Seleucid calendars.

For ease in practical computation, as the author says, all mean positions are given, to seconds (sometimes thirds) of arc for years 1, 31, 61, ···, 1321 A.H. The mean rates, to the same precision, are given for 1, 2, 3, ···, 30 Hijra years; 1, 2, 3, ···, 12 Hijra months, 1, 2, 3, ···, 29 days, also for 1, 2, 3, ···, 60 days and for 1, 2, 3, ···, 12 Persian (i.e. 30-day) months.

This zīj has a good deal of material on the so-called *world-days* (*cf.* **63**). One such "mighty" period is taken to be 36×60^8 ordinary days in length. By dividing each of the planetary periods into this number the author obtains a table of the number of times each of the small cycles is contained in the large one. This is an example of the sort of thing *al-Bīrūnī* (*Chron.*, p. 25) had in mind in his criticism of the zīj of Abū Ma'shar (*cf.* above p. 133).

F and G. PLANETARY EQUATIONS AND LATITUDES

As usual.

H. STATIONS AND RETROGRADATIONS

As in §5, H.

I. PLANETARY SECTORS

None.

J. PARALLAX

This zīj, like **55**, has the *Almagest* table of solar and lunar parallax in the altitude circle.

It also has Theon's tables of the components, for all climates, and additionally, one of the same type for $\varphi = 38°$, the latitude of Marv.

K. ECLIPSE THEORY

This section has tables of $\lambda_s'(\bar{\lambda}_s)$ and $\lambda_m(a_m)$ in degrees per hour and day, $r_s(\bar{\lambda}_s)$, $r_m(a_m)$, and $r_w(a_m)$ all to seconds of arc, for each four degrees of the arguments.

There is a table of earth–moon distances at syzygies, computed to minutes of the lunar deferent radius (taken as 1, 0), for each 6° of a_m.

The Ptolemaic digital conversion table is reproduced.

There is a table of mean conjunctions and oppositions tabulated for Hijra years 1, 31, 61, ···, 1291, and motions in 1, 2, 3, ···, 30 Hijra years, and individual months. Functions given are:

$$t,\ \lambda,\ a_m,\ \bar{\lambda}_m - \lambda_n,\ \lambda_{apS},\ \text{all to seconds.}$$

There is a table of lunar eclipses which, for $\beta_m = 0; 0°, 0; 1°, 0; 2°, \cdots, 1; 5°$, gives magnitudes, times of immersion and totality, with their equations (i.e. their changes from greatest to least lunar distance), all to minutes of the units used. An interpolation scheme is tabulated for each six degrees of a_m.

A table of solar eclipses is set up as above, but with $\beta_m = 1', 2', 3, \cdots, 34'$, and giving no time of totality.

There is a table of eclipse inclination computed to minutes of arc for each digit.

L. VISIBILITY THEORY

The sections on visibility conditions in this zīj are more extensive than in any other seen by the present writer, and a general investigation of the subject might well commence with the relevant parts of this work.

There is a table of visibility limits ascribed to Thābit (**93**). In this table, for each six degrees of a_m three entries are made: (1) the arc of *total light* (*qaus an-nūr al-kulliya*), (2) the *equation*, and (3) the *extreme distance* (*ghāyat al-bu'd*), being (1) + (2), all to minutes of arc.

There follows the author's (al-Khāzinī's) version of the same function. This table specifies the limits for good, poor, and mean visibility, to minutes of arc, for a set of λ_m' ranging from 12; 6° to 14; 27° per day.

There is a table of *Difference of Risings in the Climates* (*Ikhtilāf al-Maṭāli' fī'l-Aqālīm*), in two versions, one "according to the opinion of al-Battānī" (**55**), the other "according to the first opinion" (i.e., that of the ancients?). The argument has the range 0;0, 0;10, 0;20, ···, 13;40, and entries are to minutes, with an interpolation scheme for each six degrees of a_m.

Tables of planetary arcs of visibility are computed to minutes of arc as a function of the planet, climate and zodiacal sign. None of the entries are those of the *Almagest*. These tables are taken over, in part, by zījes **6**, **12**, and **20**, and may have been worked out by al-Khāzinī himself.

Auxiliary to these are interpolation schemes for the visibility arcs of the superior planets, for each ten degrees of the anomaly.

M. GEOGRAPHICAL POSITIONS

None.

N. STAR TABLES

There is a table of the latitude, longitude (for 500 A.H.), temperaments, and magnitudes of forty-six stars. Coordinates terminate in multiples of ten minutes, and are those of the *Almagest*, the longitudes having an added constant for precession.

O. ASTROLOGICAL TABLES

This zīj has a table said to be of Bīrūnī's method of *projections of the rays*, taking latitudes into consideration. There are two entries for each value of the argument, $\beta = 0;30°, 1;0°, 1;30°, \cdots, 10;0°$. The first is said to be "for the ray" (*Lish-shu'ā'*), and is in the vicinity of sixty, the second is "for the latitude" and ranges from zero to 4; 58, 51.

There is a table of the *Equation of the Anniversary Mean*. It is the function

$$2;12,23 \sin\theta = \max e_s \sin\theta$$

tabulated for each integer degree of θ, here the argument of the solar anomaly, to seconds of arc, with tabular differences.

There is also a table of the *Equation of the Excess of Revolution*, the function

$$0;13,40 \cos\theta$$

tabulated like the preceding function and with the same argument, but without tabular differences.

There are tables of the motion of various kinds of apheses, computed to minutes of arc, for 1, 2, 3, $\cdots$, 12 (Persian, i.e. 30-day) months. The categories are two-fold: (1) of the *burj al-muntihā'*, three kinds, those progressing in the course of a year through one sign, thirteen signs, and 169 signs; (2) of the *month and year indicators*, three kinds, those which in the course of a year progress around one revolution, thirteen signs, and twelve revolutions.

There is a table of the initial times of the *cosmic years* (*sinī al-'ālam*, i.e. vernal equinox) to seconds, for 1, 2, 3, $\cdots$, 20 years, and scores of years from 1386 to 1906 of the era of Alexander and from 444 to 964 of Yazdigird.

There is a table of the position of *al-Kaid* (*The Comet*?, *cf.* §4, O) carried to two fractional places, the motion of which passes through one sign in twenty-four years. The argument of the tables is 1, 2, 3, $\cdots$, 12, 24, 36, $\cdots$, 144 years.

13. ABSTRACT OF 6, THE *ZĪJ-I ĪLKHĀNĪ* OF *NAṢĪR AD-DĪN AṬ-ṬŪSĪ*, *c.* 1240

A. CHRONOLOGY

This zīj has material on the Hijra, Yazdigird, Seleucid, Jewish, Malikī, and Chinese-Uighur calendars.

B. TRIGONOMETRIC FUNCTIONS

There are tables of:

Sin θ, to three places, for $\theta = 0;0°, 0;1°, 0;2°, \cdots, 90;0°$,

Vers θ, to three places, for $\theta = 0;0°, 0;1°, 0;2°, \cdots, 180;0°$,

$$\text{Tan}\,\theta, \text{ for } \theta = \begin{cases} 0;0°, 0;1°, 0;2°, \cdots, 45;0°, \text{ to three places,} \\ 45;10°, 45;20°, 45;30°, \cdots, 89;0°, \text{ to four places,} \\ 89;10°, 89;20°, 89;30°, \cdots, 89;50°, \text{ to five places,} \end{cases}$$

$$\left.\begin{array}{l} 12 \cot\theta, \text{ to three places,} \\ 7 \cot\theta, \text{ to two places} \end{array}\right\} \theta = 1°, 2°, 3°, \cdots, 90°.$$

C. SPHERICAL ASTRONOMICAL FUNCTIONS

There are also tables of $\delta_1(\theta)$ and $\delta_2(\theta)$, to seconds of arc, for $\theta = 0;0°, 0;6°, 0;12°, \cdots, 360;0°$ for $\epsilon = 23;30°$.

Tables of ascensions, all computed to two fractional places and for each integer degree of λ are:

$$A_0(\lambda) + 90°, \quad \text{and} \quad A_\varphi(\lambda),$$

for $\varphi = 1°, 2°, 3°, \cdots, 53°$, also for $\varphi = 37;20°$, for Marāgha the site of Naṣīr ad-Dīn's observatory.

There are tables giving, for $\varphi = 37;20°$ (Marāgha), and for each integer degree of the argument,

$D(\lambda_s)$, to seconds of time,

and $\max h_s(\lambda_s)$ to minutes of arc.

Another table, of a type not seen in any other zīj, gives the hours from sunrise (or to sunset) to seconds, and the equivalent equatorial angle, to seconds of arc, as a function of h_s and $\max h_s$ for the day in question, for $\varphi = 37;20°$ and each integer degree of both arguments.

D. EQUATION OF TIME

There is a table, computed to two places, of $E(\lambda_s)$, for each integer degree of λ_s.

Another table, computed to the same precision as the one above, and for the same domain of the independent variable, gives the change in the moon's mean longitude to be allowed for on account of the equation of time.

E. MEAN MOTIONS

All mean positions, including the apogees, are given for years 601, 602, 603, $\cdots$, 701 of Yazdigird. In the copy used, a second table of the same character but for years 861, 862, 863, $\cdots$, 890, was also present, apparently added by a user who lived after the time of Naṣīr ad-Dīn. Motion of the means is also given, for 1, 2, 3, $\cdots$, 10, 20, 30, $\cdots$, 100, 200, 300, $\cdots$, 1200 Yazdigird (Egyptian) years; 1, 2, 3, $\cdots$, 12 months; 1, 2, 3, $\cdots$, 30 days; 1, 2, 3, $\cdots$, 24 hours; and for change of terrestrial longitude. All the above tables are computed to seconds of arc.

Precessional motion is given to four significant places for complete Persian (i.e. Egyptian) years, an approximation to the basic rate being 0; 0, 51, 25, 43° per Egyptian year.

F. PLANETARY EQUATIONS

As usual, except that a third lunar equation is tabulated, being the small effect on λ caused by the fact that the inclined orbit of the moon is not in the ecliptic plane, but inclined to it at five degrees.

The tables are computed for an interval of each twelve minutes instead of each degree.

G. PLANETARY LATITUDES

As usual, except that β_m is computed for every twelve minutes of the argument.

H. STATIONS

As in §5, H.

I. PLANETARY SECTORS

There are tables of deferent and epicycle sectors computed to degrees only.

J. PARALLAX

The Theon parallax component tables for the third, fourth, and fifth climates are present, together with another of the same type for $\varphi = 38°$. The entries of the latter are identical with those of al-Khāzinī (**27**, §12, J) and may be taken from his zīj, for Naṣīr ad-Dīn's latitude for his observatory, Marāgha, is given as 37; 20°.

K. ECLIPSE THEORY

There is a table of lunar eclipses in which the arguments are:

$$\beta_m = 1', 2', 3', \cdots, 67', \quad \text{and}$$
$$\lambda_m' = 11;36°, 11;48°, 12;0°, \cdots, 15;0° \text{ per day.}$$

Functions obtained are magnitude, time of immersion, and time of totality, all to two places.

A table of solar eclipses has the same arguments, but the apparent β_m extends to 35′ only. The table gives magnitude in diametral and areal digits, and time of immersion, all to two places.

L. VISIBILITY CONDITIONS

There is a table of the arc of visibility, the same as the table described in §12, *L*, but for the fourth climate only.

M. GEOGRAPHICAL POSITIONS

This zīj has a table giving the latitudes and longitudes of about 245 localities, the coordinates terminating in multiples of five minutes. For thirty-five localities the length of the longest day is also given.

N. STAR TABLES

There is a table of the latitude, longitude, temperaments, and magnitudes of sixty stars. Coordinates terminate in multiples of five minutes.

Another table gives only eighteen stars but giving the coordinates (λ, β) as observed by Ptolemy, Ibn al-Aʿlam (**70**), and Ibn Yūnis (**14**) as well as those of Naṣīr ad-Dīn himself. See §17 below. By comparison of the latitudes reported it is clear that, for these stars at least, independent observations of stellar positions were made at Marāgha.

O. ASTROLOGICAL TABLES

There is a table of *equalization of the houses* for each degree of the ecliptic, to minutes of arc.

There is a table of the *excess of revolution*, for 1, 2, 3, $\cdots$, 12 years, to minutes of arc.

Of *nativity* and *world* periods there are substantially the same tables as those noted in the Khāqānī Zīj (**20**, §15, O below), except that the table of positions for the world periods runs through the Malikī years 181, 182, 183, $\cdots$, 300.

P. MISCELLANEOUS

This zīj, like **20** (in §15, P) and **12** (in §16, *P*), has three tables of the function

$$\frac{24d}{\lambda'},$$

computed to seconds of time.

In the first the arguments have the domain

$$d = 1', 2', 3', \cdots, 60', \quad \text{and}$$
$$\lambda_s' = 57', 58', 59', \cdots, 62' \text{ per day.}$$

Thus the table gives the number of hours required for the sun to traverse arcs of longitude of d minutes at a rate of λ'.

The second performs the same function for the moon, the arguments running

$$d = 1°, 2°, 3°, \cdots, 10°, 20°, 30°, 40°, 50°, \quad \text{and}$$
$$\lambda_m' = 9;30°, 9;31°, 9;32°, \cdots, 16;10° \text{ per day.}$$

The third predicts the time of conjunction of two planets whose difference in longitude is d and which are approaching at the rate of λ'. The domain of d is as that for the moon above, and

$$\lambda' = 9;36°, 9;39°, 9;42°, \cdots, 17;9° \text{ per day.}$$

14. ABSTRACT OF 11, *AZ-ZĪJ AL-JADĪD* OF *IBN ASH-SHĀṬIR*, *c.* 1350

A. CHRONOLOGICAL

This zīj has material on the Hijra, Yazdigird, Seleucid, and Coptic calendars, together with a table of the lunar mansions.

B. TRIGONOMETRIC FUNCTIONS

There are tables of:

Sin θ, to four places, for $\theta = 0°, 1°, 2°, \cdots, 90°$, with tabular differences,

Tan θ, to three places, for $\theta = 0;0°, 0;30°, 1;0°, \cdots, 89;30°$, with tabular differences,

$\left.\begin{array}{r}12 \cot \theta,\\ 7 \cot \theta,\end{array}\right\}$ to three places, for $\theta = 1°, 2°, 3°, \cdots, 90°$,

and arc cot $\frac{d}{12}$, to three places, $d = 0;0, 0;15, 0;30, \cdots, 60;0$ digits.

C. SPHERICAL ASTRONOMICAL FUNCTIONS

There are also tables of:

$\delta(\theta)$, to seconds of arc, for each integer θ, with tabular differences, for $\epsilon = 23;31°$.

Tables of ascensions include,

$A_0(\lambda), A_0(\lambda) + 90°, A_{33;30°}(\lambda)$, computed to seconds of arc, for each integer degree of λ, and having tabular differences; also

$A_\varphi(\lambda)$, to minutes of arc, $\lambda = 0°, 6°, 12°, \cdots, 360°$, $\varphi = 1°, 2°, 3°, \cdots, 51°$.

In the ascension tables the value $\epsilon = 23;30°$ is used.

There are tables, for $\varphi = 33;30°$ (Damascus), and for each integer degree of the argument, giving

$D(\lambda_s)$ to seconds of time,

and $\frac{D(\lambda_s)}{2}$ to minutes of arc.

With the latter table is a column for tabular differences, left blank in the manuscript used.

There are tables, for each integer degree of the argument, of

$$\bar{h}_s(\lambda_s), \quad \text{Tan } h_s(\lambda_s), \quad \text{and} \quad \text{Cos } h_s(\lambda_s),$$

all for $\varphi = 33;30°$ (Damascus) (*cf.* §5, C).

Ibn ash-Shāṭir has made a table of 184 trigonometric and spherical-astronomic identities, each identity expressed as the four terms of a proportion. The first of the list, for example, is equivalent to the expression

$$\frac{1,0}{\text{Tan } \varphi} = \frac{\text{Tan } \delta(\lambda_s)}{\text{Sin } \frac{\Delta D}{2}}.$$

D. EQUATION OF TIME

There is a table of $E(\lambda_s)$, computed to seconds for each integer degree of the argument.

E. MEAN MOTIONS

Mean positions, including $\bar{\lambda}_s, \bar{\lambda}_m, a_m, \lambda_n, \bar{\lambda}_s$ less apsidal motion, apsidal motion separately, $\bar{\lambda}$ for the superior planets, and a for the inferior planets, are given at epoch of the Hijra calendar and at years 30, 60, 90, ⋯, 900 A.H. for the original zīj and to 1050 A.H. added by a later hand. The motions of these in 1, 2, 3, ⋯, 30 Hijra years, and in the appropriate months, days, hours, and minutes are displayed, mostly to thirds of arc.

Precessional motion is tabulated down to months and days, to four fractional places. An approximation to the basic parameter is $0;0,0,8,27,14°$ per day.

F. PLANETARY EQUATIONS

As usual, but with the third lunar equation (*cf.* §13, F).

G. PLANETARY LATITUDES

As usual.

H. STATIONS

There is a table of stations, computed to seconds of arc, for each degree of the argument.

I. PLANETARY SECTORS

None.

J. PARALLAX

Ibn ash-Shāṭir gives the *Almagest* table of solar and lunar parallax in the altitude circles.

There is also a tabulated interpolation scheme for modifying the above, in the case of the moon, for variation of the anomaly.

The Theon component tables for all climates are reproduced. An auxiliary table to this, also from the *Handy Tables*, and appearing in no other zīj examined, is for modifying the entries according to the lunar anomaly.

Ibn ash-Shāṭir gives his own table of P_β and P_λ computed for his location, Damascus. Differently laid out from those of Theon, the arguments are: (1) each minute of arc after sunrise, and (2) conjunctions located at the initial point of each zodiacal sign. Entries are to one place only.

There is an interpolation scheme, a set of coefficients expressed as a function of the lunar daily rate, for modifying the entries of the preceding table.

A second interpolation arrangement permits of modifying the entries of the Damascus P_β and P_λ table to allow for changes of φ to other locations.

As a completely alternative approach to the problem, the same P_β and P_λ tables appearing in the Khwārizmī Zīj (§6, J) are also given in this one, but with fuller and more accurate directions for their use.

K. ECLIPSE THEORY

There are tables of $2r_s$ and $(r_m + r_u)$ to two places for each degree of λ_s and a_m respectively.

The same functions are tabulated as two-place functions of the solar and lunar rates, the ranges being

$\lambda_s' = 57;15', 57;25', 57;35', \cdots, 61;35'$ per day, and
$\lambda_m' = 11;30°, 11;40°, 11;50°, \cdots, 15;0°$ per day.

There is a table of eclipse magnitudes defined as the function $12b/2r_m$, where the variables

$$b \equiv r_m + r_u - \beta_m = 1', 2', 3', \cdots, 36', \quad \text{and}$$
$$2r = 28', 29', 30', \cdots, 37'$$

run through the values shown.

This zīj reproduces the *Almagest* digital conversion table.

The table of lunar eclipses has as arguments:

$$\lambda_m' = 12°, 13°, 14°, 15° \text{ per day, and}$$
$$\beta_m = 1', 2', 3', \cdots, 63',$$

in terms of which duration of immersion and totality are tabulated, to two places.

The table of solar eclipses is arranged as above except that apparent β_m ranges to 36′ only, and only the time of immersion is tabulated.

L. VISIBILITY TABLES

This zīj has two lunar ripeness tables. The first is the function $(g^2 + \beta_m^2)$, computed to two places, where g, the elongation, $= 5;0°, 5;30°, 6;0°, \cdots, 21;30°$ and $\beta_m = 0;0°, 0;30°, 1;0°, \cdots, 5;0°$. This utilizes the fact that the amount of illumination received from the moon is proportional to the distance on the celestial sphere from sun to moon.

The second is of the same type except that the results have been converted into digits such that full moon is twelve digits of illumination. The range of the arguments is now

$$g = 1°, 2°, 3°, \cdots, 17°$$
$$\beta_m = 0;10°, 0;20°, 0;30°, \cdots, 5;0°.$$

For the planets this zīj reproduces the Almagest tables, except for copyists' variants.

There is a table of variation of the arc of visibility,

$$V = 14 - 6L$$

where $L = 0;35, 0;36, 0;37, \cdots, 1;22$ is the light.

There is a table of planetary *difference of setting* for the latitude of Damascus, computed to minutes of arc, for each five degrees around the ecliptic and for $\beta = 0°, 1°, 2°, \cdots, 7°$.

There is a table of the *arc of tarrying* (*qaus al-makth*) computed to seconds of arc for each degree of the argument, for the latitudes of Damascus and of Cairo.

There is a table of the *difference of rising and setting*, for all seven climates and at intervals of three degrees of the argument, computed to minutes of arc.

M. GEOGRAPHICAL POSITIONS

For about 290 cities, the latitude, longitude, and direction of the *qibla* (the azimuth of Mecca) are given, all to two places.

N. STAR TABLES

The equatorial coordinates (α, δ) of some eighty stars are given, to minutes. *Knobel* (*Chron.*, p. 16) states that this table is derived from the star table of Ibn Yūnis (**14**). He says further that the right ascensions are reckoned from the first point of Capricorn, and that the epoch of the table is 1480 (?).

O. ASTROLOGICAL TABLES

The only strictly astrological table in this zīj is one of lots, of fortune, kingship, absence, victory, gold, etc., used in the *year-transfer*. The table indicates whether the lot is to be determined from the ascendant or from some planet.

15. ABSTRACT OF 20, THE *ZĪJ-I KHĀQĀNĪ*, OF *JAMSHĪD AL-KĀSHĪ*, *c.* 1420

(Folio references are to the India Office copy)

The introduction to this zīj contains a determination, described in full detail, of the lunar mean and anomalistic motion. It is based on three eclipses observed in Kāshān by Jamshīd, and on three other eclipses, observed by Ptolemy and reported in the *Almagest*. The results obtained are

$$\bar{\lambda}_m' = 13;10, 35, 1, 52, 47, 50, 50° \text{ per day}$$
$$a_m' = 13;3, 53, 56, 30, 37, 20° \text{ per day.}$$

A. CHRONOLOGY

Full explanation and tables (ff. 6v–23r) are provided for the Hijra, Yazdigird, Seleucid, Malikī, Chinese-Uighur, and Īlkhānī calendars.

B. TRIGONOMETRIC FUNCTIONS

There are tables (ff. 35r–42r) of

Sin θ, and
Tan θ, to four places, for each minute of arc.

C. SPHERICAL ASTRONOMICAL FUNCTIONS

There are also tables (ff. 42v–44r) of

$\delta_1(\theta)$, for $\theta = 0;0°, 0;6°, 0;12°, \cdots, 360;0°$, and
$\delta_2(\theta)$, for $\theta = 0;0°, 0;12°, 0;24°, \cdots, 360;0°$, both to seconds of arc and with $\epsilon = 23;30°$.

Tables of ascensions (ff. 44v–72r, 158r) computed to seconds of arc, for each integer degree of λ, are for

$A_0(\lambda) + 90°$, and
$A_\varphi(\lambda)$, for $\varphi = 0°, 1°, 2°, \cdots, 61°, 66;30°, 75°$.

There is a table (f. 126v), computed to two places, of $E(\lambda_s)$, for each integer degree of the argument, for the year 712 Yazdigird. The motion in 100 years and 1000 years is also given.

There are also tables (ff. 132r, 133v), computed to two places for each degree of the solar mean longitude, showing the changes in mean longitude of the sun and moon required to compensate for the equation of time.

E. MEAN MOTIONS (ff. 127v–130v)

Positions of $\bar{\lambda}_s$, λ_m, $\bar{\lambda}_{apS}$ are given to thirds of arc, of a_m, λ_n, λ of the superior planets, a of the inferior planets, and λ_{ap} of all planets to seconds for years 781, 782, 783, $\cdots$, 791 Yazdigird. Motions of all of these, to the same precision, are given for 10, 20, 30, ···, 100 Yazdigird years, and appropriate numbers of months, days, and hours.

Precessional motion is tabulated (f. 166v) down to months, to three fractional places. The parameter is taken over from the Īlkhānī Zīj (6).

F and G. PLANETARY EQUATIONS AND LATITUDES

There is a complete set of tables (ff. 131r–140v) of the usual sort, arranged as in §13, F and G. In addition to this there are extensive tables (ff. 142r–155r) for simplifying the computation of planetary latitudes and longitudes (*cf.* **88**). In view of what is known of the rest of al-Kāshī's work it is not probable that there is any essential divergence in the latter from the Ptolemaic theory, but they may be of interest from the computational point of view.

H. STATIONS AND RETROGRADATIONS (f. 141v)

As in §5, H.

This zīj has tables showing maximum and minimum duration of forward motions and retrogradations for each planet, expressed both in days (carried to hours) and degrees (carried to seconds).

I. PLANETARY SECTORS

There is a table (f. 141r) of epicycle and deferent sectors, computed both for velocity and distance, to minutes of arc.

J. PARALLAX

Kāshī gives a table (f. 163r) of adjusted lunar parallax (at syzygy) in the altitude circle, computed to two places for each integer degree of zenith distance. He also has an interpolation scheme for modifying the tabular values with variation of the lunar anomaly.

There are table lay-outs (ff. 164v–165r) for P_λ and P_β, not for the latitudes of the seven climates, but for $\varphi = 20°$, $30°$, $40°$, and $50°$. But in both available copies of the zīj only the table for $\varphi = 30°$ has been filled in. The actual entries only crudely approximate the accurate values of, say, Ulugh Beg (*cf.* §16, J), and the present writer is thus far at a loss to explain the mode of computation.

The third type of table (f. 185r) purports to give the horizontal parallax of Venus, the sun, and the moon, as functions of distance.

K. ECLIPSE TABLES

There are tables (f. 163r) of $\lambda_s'(\bar{\lambda}_s)$ and $\lambda_m'(a_m)$, both in the ecliptic and in the plane of the orbit, for each five degrees of the arguments, to two places.

There are also tables (f. 163r) of $r_s(\bar{\lambda}_s)$, $r_m(a_m)$, and $r_w(a_s, \bar{\lambda}_m)$, for each 30° of $\bar{\lambda}_s$ and each five degrees of a_m, all computed to two places.

A table (f. 162v) of mean conjunctions and oppositions gives positions in years 801, 802, 803, ···, 811 A.H., and motions for 10, 20, 30, ···, 100, 200, 300, ···, 1000 Hijra years, and mean lunar months. Entries are t, λ, and λ_{ops} to four fractional places of time or arc, and a_m and λ_n to seconds of arc, for the conjunction or opposition in question.

The table (f. 163v) of lunar eclipses has arguments

$\beta_m = 2', 4', 6', \cdots, 70'$, and
$\lambda_m' = 11;50°, 12;10°, 12;30°, \cdots, 14;50°$ per day.

Functions tabulated are magnitude, in diametral and areal digits, time of immersion and time of totality, all to two places.

The table (f. 165v) of solar eclipses is arranged as above, except that β_m ranges only to 34′ and the duration of totality is not tabulated.

L. VISIBILITY TABLES (f. 166v)

There is a table of the lunar *equation of setting* computed to minutes of arc for each zodiacal sign, each integer β_m, and for the second, third, fourth, and fifth climates.

For the planets there are tables of the *arc of visibility*, as in the Sanjarī Zīj (§12, L) and with the same values, but for the third and fourth climates only.

M. GEOGRAPHICAL POSITIONS

The latitude and longitude (from the Fortunate Isles) of 516 localities is given (ff. 72v–74v), the coordinates terminating in multiples of five minutes. The arrangement is by climates, as follows:

Equator to the First Climate,	25
In the First Climate,	36
In the Second Climate,	48
In the Third Climate,	118
In the Fourth Climate,	158
In the Fifth Climate,	78
In the Sixth Climate,	30
In the Seventh Climate,	14
North of the Seventh Climate,	9

There is in addition a table (f. 74v) showing those values of φ which bound the climates and those which mark their middles.

N. STAR TABLE

The latitude, longitude, magnitude, and temperament of eighty-four stars are given (f. 157r). The author states the coordinates have been obtained either from the *Almagest* or from the Īlkhānī Zīj (6), with correction for precession.

O. ASTROLOGICAL TABLES

The following tables (ff. 205v–209r) of periods are relative to *nativities*, all being tabulated for arguments of 1, 2, 3, ···, 365 days, and 1, 2, 3, ···, 24 hours:

The Aphesis of Birth Indicators (Persian: *Tasyīr-i Dalā'il-i Aṣlī*) which progress a degree per year, computed to seconds of arc. (The inverse of this function is also tabulated, the days and hours corresponding to 1, 2, 3, ···, 60 [*tasyīr*] minutes and seconds being given.)

The *Aphesis of the Anniversary* (Persian: *Tasyīr-i Taḥvīl*) progresses twelve signs per year, tabulated to minutes of arc.

The *Aphesis of the Anniversary Centers* (*Autād*) progressing a revolution plus 87;15° per year, is tabulated to minutes of arc.

The *Annual Progression* (Persian: *Intihā'-i Sinavī*) goes through a sign per year, and is tabulated to seconds of arc.

The *Monthly Progression* moves thirteen signs per year, and is tabulated to minutes of arc.

There is a table of the *firdaria* and its planetary associates, for a cycle of 75 years, for diurnal and nocturnal nativities.

The table layout for *cycles* has been left blank in the zīj.

There follows an analogous set of tables (ff. 209v–212v) relative to periods of the *world*:

For an argument consisting of years 301, 302, 303, ···, 402 Malikī, and 100, 200, 300, ···, 1000, 2000, 3000, ···, 10,000, 20,000, 30,000, ···, 100,000 years the zodiacal positions of the following are shown: the four (*mighty*, *large*, *middle*, and *small*) apheses (moving by a degree in 1000, 100, 10, and 1 years respectively) and the *mighty firdaria* (with period of 360 × 7 × 12 years).

There is a table locating eighty-four years of the *mighty firdaria* in the Malikī calendar, together with the associated signs and planets.

There is a table of the motion of the *mighty firdaria*, to minutes, for 1, 2, 3, ···, 60, 120, 180, ···, 360 years.

The motion of the *big firdaria* is tabulated to minutes for 1, 2, 3, ···, 78 years.

There is a table of a cycle of the *middle* and *small firdaria* with their associates.

P. MISCELLANEOUS

This zīj has a table (f. 157r) of distances of all the planets from the center of the earth. Distances are given in sixtieths of the deferent radius to two (or three) places, with the epicycle center assumed at deferent apogee, for each five degrees of the anomaly. Another column gives for each entry the equation, the amount by which the entry must be decreased to obtain the corresponding distance when the epicycle center is at perigee. Finally there is an interpolation arrangement for modifying the equation for general positions on the deferent, tabulated for each five degrees from apogee.

There are two tables (ff. 158v–162r) of the function m/λ', computed to three places. Domains of the variables are

$m = 1', 2', 3', \cdots, 10', 20', 30', 40', 50'$, and
$\lambda' = 2;23', 2;24', 2;25', \cdots, 2;34'$ per hour, also
$\lambda' = 23;45', 23;50', 23;55', \cdots, 42;25'$ per hour.

For the application of this table see §13, P.

16. ABSTRACT OF 12, THE *ZĪJ-I SULṬĀNĪ*, OF *ULUGH BEG*, c. 1440

A. CHRONOLOGY

Explanation and tables are provided for the Hijra, Yazdigird, Seleucid, Malikī, and Chinese-Uighur calendars.

B. TRIGONOMETRIC FUNCTIONS

There are tables of:

Sin θ, to five places, $\theta = 0;0°, 0;1°, 0;2°, \cdots, 90;0°$, with tabular differences,

Tan θ, to five places, $\theta = 0;0°, 0;1°, 0;2°, \cdots, 45;0°, 45;10°, 45;20°, \cdots, 79;50°$, with tabular differences

$\left.\begin{matrix} 12 \cot \theta, \\ 7 \cot \theta, \end{matrix}\right\}$ to four places, $\theta = 1°, 2°, 3°, \cdots, 90°$.

The tables for Sin θ and Tan θ were partially published by *Schoy* (*Mas'ūdī*).

C. SPHERICAL ASTRONOMICAL FUNCTIONS

There are also tables computed to three fractional places, of:

$\delta_1(\theta)$, for $\theta = 0;0°, 0;3°, 0;6°, \cdots, 360;0°$,
and $\delta_2(\theta)$, for $\theta = 0;0°, 0;6°, 0;12°, \cdots, 360;0°$.

Tables of ascensions, all computed to three fractional places except where noted, and for each integer degree of λ, are

for $A_0(\lambda) + 90°$, with tabular differences,
and $A_\varphi(\lambda)$ for $\varphi = 39;37,23°$ (that of Samarqand,

the site of Ulugh Beg's observatory) to two fractional places, with tabular differences,

for $\varphi = 0°$, with tabular differences,
for $\varphi = 1°, 2°, 3°, \cdots, 50°$.

D. EQUATION OF TIME

There is a table, computed to three fractional places, of $E(\lambda_s)$, for each integer degree of the argument.

E. MEAN MOTIONS

All mean positions, including solar apsidal longitudes, are shown for Hijra years 841, 842, 843, ···, 871. Motions of these in 30, 60, 90, ···, 300, 600, 900, ···, 1200 Hijra years and appropriate months, days, hours and minutes are also given. There is a table for determining the effect on mean position of change in terrestrial longitude. All the above are to seconds of arc, except the solar and lunar parameters, which are to thirds.

F and G. PLANETARY EQUATIONS AND LATITUDES

As usual, but with e_s computed to thirds of arc, and with a table of the third lunar equation.

H. STATIONS AND RETROGRADATIONS

As in §5, H.

I. PLANETARY SECTORS

There is a table of sectors, of all four categories, computed to seconds of arc.

J. PARALLAX

There is a table of solar parallax in the altitude circle, computed to three significant places for each degree of zenith distance.

Another table gives the adjusted lunar parallax (at syzygy) in the altitude circle, also computed to three significant places, and for the same values of the argument. An interpolation scheme makes it possible to modify the entries for variation of the lunar anomaly.

There is a set of component tables of the Theon type, but computed for $\varphi = 25°, 30°, 35°, \cdots, 50°$, to seconds of arc for the same ranges of arguments in the individual tables as with Theon.

K. ECLIPSE THEORY

There is a table of r_s for each degree of mean distance from apogee, of r_m and r_u for each five degrees of a_m, all to two places.

A table of lunar eclipses has the arguments:

apparent $\beta_m = 0', 2', 4', \cdots, 64'$, and $\lambda_m = 11;50°, 12;10°, 12;30°, \cdots, 14;50°$ per day.

Functions tabulated are: magnitude in diametral and areal digits, and duration of immersion and totality, all to two places.

For solar eclipses the layout is as above, except that β_m goes only to 34′.

L. VISIBILITY TABLES

There is a table of the *lunar equation of setting*, computed to minutes of arc for

$$\beta_m = 1°, 2°, 3°, 4°, 5°,$$
$$\varphi = 25°, 30°, 35°, 40°,$$

and for each zodiacal sign.

The tables of planetary arc of visibility which appear in the Sanjarī Zīj (§12, L) are reproduced here also, but for the third and fourth climates only.

M. GEOGRAPHICAL POSITIONS

Latitudes and longitudes (from the Fortunate Isles) of some 240 towns are tabulated, to minutes of arc. The arrangement is by regions rather than by climates.

N. STAR TABLE

As remarked above, the star table of this zīj has been published, most recently by *Knobel.* It gives the ecliptic coordinates (β, λ) to minutes, of 1018 stars. Most of these are the results of independent observation, although some have been obtained from the *Almagest* table with the longitudes corrected for precession.

O. ASTROLOGICAL

Of periods relating to *nativities* and to the *world* there are substantially the same tables as in the Khāqānī Zīj (**20**, §15, O).

P. MISCELLANEOUS

There is a table of greatest lunar distances, with the corresponding equation for least distances, and an interpolation scheme, computed to three places, for each degree of both arguments.

There is also a table of solar distance, to three fractional places (deferent radius = 1, 0), for each degree of distance from the apogee.

There is a table called "Proportional Parts of the Hour Angle of the Two Centers," being

$$n \times 6;6,15,25 \quad \text{and} \quad n \times 0;6,6,15$$

for $n = 1, 2, 3, \cdots, 60$. If the length of the sidereal year is taken as $6, 5; 15, 25^d$, then its length *measured in sidereal days* will be one more, or $6, 6; 15, 25^d$. This table enables the user to determine easily the number of sidereal days in, say, k sidereal years.

The same tables of $\frac{24d}{\lambda}$ as are given in **6** (§13, P above) appear also in this zīj. The ranges of the arguments are somewhat different, notably in the third table, where

$$d = 1, 2, 3, \cdots, 60, \quad \text{and}$$
$$\lambda' = 0;1, 0;2, 0;3, \cdots, 2;30.$$

17. THE DATA SUMMARIZED

Having assembled a considerable mass of evidence, we turn finally to the problem of interpretation. The easiest measure to apply is that of sheer quantity. Figure 6 is a plot locating roughly within time and space as many of the 109 listed zījes as can be fixed with any assurance. To the right of the main plot a graphical frequency distribution by centuries is displayed.

This chart can by no means be regarded as a final representation. Two or three times during the

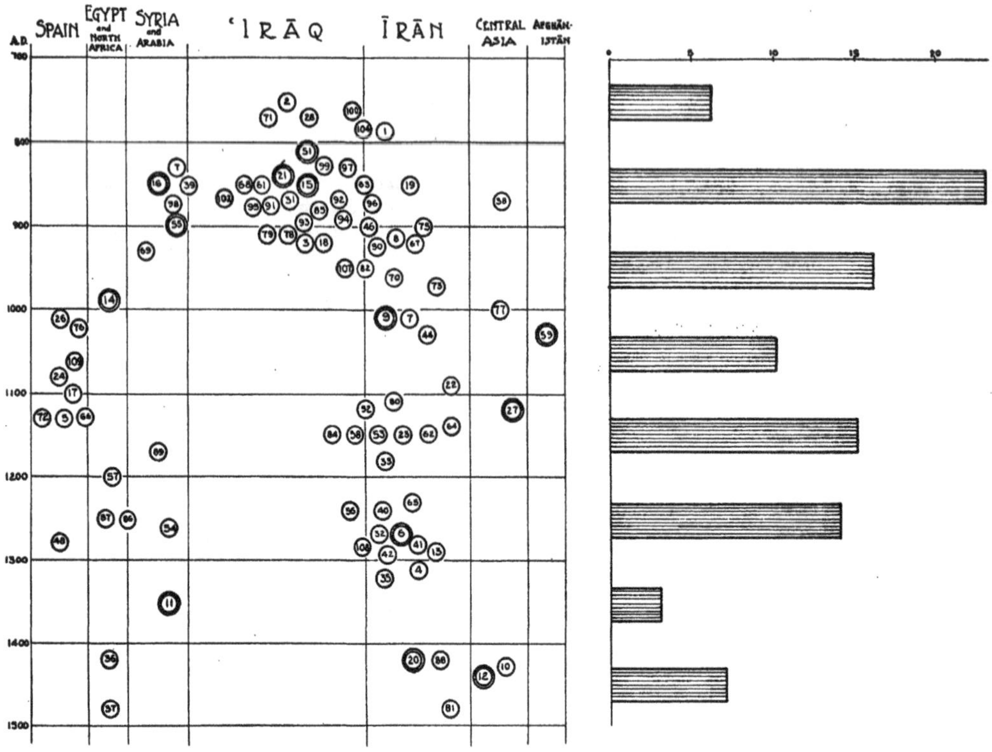

FIG. 6. Distribution of zījes by geographical location and century.

preparation of the paper the discovery of a new source naming numbers of previously unlisted handbooks altered the distribution essentially. There is no reason for thinking that the possibilities have been exhausted. Additional zīj names and authors may be expected to turn up. But the picture as it stands is worth examination.

Production of tables rises sharply to a maximum in the ninth century, in coincidence with the prospering of the Abbasid caliphate in Baghdād. Thereafter activity falls off more or less steadily, although it is impressive at all times during the period considered, for seven zījes are known to have been produced during the fifteenth century, the last one included. Geographically, by far the thickest cluster of activity is at Baghdād, the later dispersion of scientific effort reflecting dispersion of political power. As time passes the locus of centroids veers eastward to Iran, although a scattering of activity maintains itself in Spain and North Africa.

Zījes abstracted are indicated by doubled circles. It will be noted that all centuries of the period are represented in one or more abstracts. From the point of view of geographical distribution, however, it is unfortunate that no handbook written in Spain has been abstracted.

Having dealt with quantity and distribution, two related inquiries may next be made; to what extent can groups of interdependent zījes be identified and to what degree are they the results of independent observation.

Any definitive setting up of table families is contingent upon the assembling and study of additional hundreds of numerical parameters, e.g., mean motions. But on the basis of what is immediately at hand, several groups can be noted:

Zījes of the Sindhind family are enumerated under **28**.

The Battānī Zīj (**55**) is the basis of **7**, **9**, **44**, **49**, and **65**.

There is a group of related Spanish Arabic zījes including **5**, **24**, **48**, **66**, and **72**.

The zījes of al-Fahhād, **23**, **53**, **58**, **62**, **64**, and **84**, and that of al-Fārisī, **54**, are to be grouped together.

The complicated relations between the members of the Abū al-Wafā' family are best seen from figure 7. A number of basic mean motions in degrees per day and attributed to Abū al-Wafā' have been found in the Berlin copy of the Ḥabash Zīj (**15**). These are reproduced below for future reference in attempts to classify additional zījes:

☉ (longitude)	0; 59, 8, 20, 43, 17, 38, 41, 42, 25, on f. 29,
☾ (longitude)	13; 10, 35, 1, 55, 37, 39, 6, 16, 45, 43, on f. 33,
☾ (anomaly)	13; 3, 53, 56, 17, 50, 15, 50, 59, 17, 31, on f. 34,
♃ (longitude)	0; 4, 59, 16, 58, 50, 44, 30, 49, 53, 17, on f. 45,
♀ (anomaly)	0; 36, 59, 29, 7, 49, 1, 36, 9, 21, 59, on f. 54,
☿ (anomaly)	3; 6, 24, 6, 55, 45, 22, 0, 37, 26, 24, on f. 57.

In discussing Islamic observational activity several general remarks should first be made. Although astronomical observations were more precise and far more numerous than during any preceding period, no zīj was based on newly observed values for all parameters, and most observers contented themselves with a few, easily observed phenomena like the solar apogee and equation. This was partly because observation cannot be completely divorced from theory, and the Ptolemaic system was on the whole adequate for the instruments of the time. The genius of Ptolemy is shown nowhere to better advantage than in his planetary latitude theory, and this was left severely alone by the Islamic astronomers. Here, only if the basic theory had been overhauled would new observations have been of much utility.

With these reservations as to scope, the following individuals or groups can safely be asserted to have made independent observations:

1. Al-Nihāwandī (**1**) at Jundīshāpūr.
2. The "Companions of the Mumtaḥan," including Yaḥyā (**51**), al-Jawharī (**99**), Sanad ibn 'Alī (**96**), al-Marvarūdī (**97**), and Ḥabash (**15**) at Baghdād and Damascus.
3. Ad-Dīnawarī (**19**) at Isfahān.
4. The Banī Mūsā (**91**, **92**).
5. Al-Māhānī (**98**) at Baghdād.
6. Thābit (**93**) at Baghdād.
7. As-Samarqandī (**38**) at Samarqand.
8. Al-Battānī (**55**) at Raqqa.
9. The Banī Amājūr (**8**, etc.) at Baghdād and Shīrāz.
10. Aṣ-Ṣūfī (**107**) at Baghdād.
11. Ibn al-A'lam (**70**) at Baghdād.
12. Abū al-Wafā' (**73**) at Baghdād.
13. Ibn Yūnis (**14**) at Cairo.
14. Al-Khujandī (**60**) at Rayy.
15. Al-Bīrūnī (**59**) at Ghazna and Khwārizm.
16. The Toledan Observations (**24**).
17. Al-Khāzinī (**27**) at Marv.
18. Al-Fahhād (**23**, etc.).

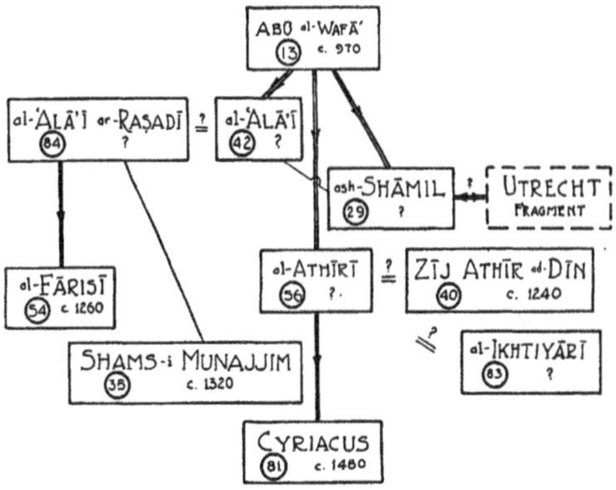

FIG. 7

19. The Īlkhānī Observations (**6**) at Marāgha.
20. Ibn ash-Shāṭir (**11**) at Damascus.
21. Al-Kāshī (**20**) at Kāshān.
22. Ulugh Beg (**12**) at Samarqand.

In connection with observational astronomy, table 1 shows the ecliptic coordinates of eleven prominent stars as observed by four Muslim astronomers. The coordinates found by Ibn al-A'lam, Ibn Yūnis, and Naṣīr ad-Dīn were all excerpted from **6**. The Mumtaḥan values were obtained from both **16** and **51**. The *Almagest* coordinates for the same stars are also displayed. Here again the object is to provide a basis for the future examination of star tables in other zījes. Where corresponding latitudes in general differ from all five values shown, an independently observed table is indicated. In cases where the latitudes are the same, there should be a constant precessional difference between corresponding longitudes, and this may be used to date the table.

It has been felt useless to reproduce sets of equatorial coordinates here, since these change with precession. Nevertheless, it should be stated that the Muslim astronomers made use of right ascensions and declinations already in the ninth century (*cf.* §8, N). The earliest previously noted use of equatorial coordinates for star tables was in the thirteenth century. (See Houzeau, J. C., "Vademecum de l'astronomie," *Annales de l'observatoire royal de Bruxelles*, 1882, p. 156; Sédillot, J. J., "Traité des instruments astr. . . . ," Paris, 1834, vol. I, pp. 191, 276.)

Before proceeding to a description of trends and developments by topics, it will be useful to consider the part played by Islamic astronomy as embedded in the whole setting of ancient science. Figure 8 exhibits, graphically if crudely, the extent of the three major pre-Islamic astronomical disciplines and that of the Islamic period as rectangular spreads on a temporal-geographical net. Thickness of the rectangles is a rough indication of significance, the circular

TABLE 1. EXCERPTS FROM FIVE STAR TABLES

		PTOLEMY		MUMTAḤIN		IBN al-A'LAM		IBN YŪNIS		NAṢĪR ad-DĪN	
		λ	β	λ	β	λ	β	λ	β	λ	β
اخر النهر	θ Eridani	0^s 0;10°	−53;30°	0^s 10;10°	−53;30°	0^s 16;42°	−53;30°	0^s 16;20°	−53;28	0^s 24;15°	−51;45°
عين الثور	α Tauri	1 12;40	−5;10	——	−5;45	1 28;40	−5;15	1 29;7	−5;15	1 29;22	−5;13
عيوق	Capella	1 25;0	+22;30	2 5;5	+22;50	2 10;50	+22;50	2 11;32	+22;3	2 11;10	+22;40
الشعرى اليمانية	Sirius	2 17;40	−39;10	2 27;50	−39;20	3 3;40	−39;4	3 4;2	−39;30	3 3;50	−39;10
الشعرى الشامية	α Procyon	2 29;10	−16;10	3 9;0	−16;0	3 15;45	−16;0	3 15;12	−16;2	3 15;45	−16;5
قلب الاسد	Regulus	4 2;30	+0;10	4 13;0	+0;15	4 18;45	+0;15	4 19;10	+0;10	4 19;4	+0;17
السماك الاعزل	Spica	5 26;40	−2;0	6 6;48	−2;6	6 12;33	−2;6	6 12;58	−2;10	6 13;25	−1;52
السماك الرامح	Arcturus	5 27;0	+31;30	6 7;10	+31;12	6 12;55	+31;12	6 13;49	+31;33	6 13;0	+31;35
قلب العقرب	Antares	7 12;40	−4;0	7 22;55	−4;24	7 28;47	−4;24	7 29;7	−2;25	7 29;—	−4;10
النسر الواقع	Vega	8 17;20	+62;0	8 29;0	+61;44	9 4;45	+61;45	9 5;8	+61;55	9 4;40	+61;50
النسر الطائر	Altair	9 3;50	+29;10	9 14;18	+29;12	9 20;58	+29;14	9 20;34	+29;10	9 20;40	+29;15

vignettes characterizing basic techniques in each discipline, and arrows showing the direction of influences between areas. Of astronomy in Sasanian Iran, little more can be said than that it existed, and a dotted rectangle is shown in the proper general region.

There is no need to apologize for studying Islamic astronomy in its own right, and to do so entails acquaintance, the deeper the better, with Hellenistic, Babylonian, and Hindu astronomy. For the Muslims were profoundly influenced by all three, particularly the first. And it is clear that in the Islamic period science made no advance comparable to those marked by the Ptolemaic and Newtonian systems.

But in addition to its intrinsic interest, there is a second and perhaps more cogent reason for studying medieval astronomy, namely that its masses of source material can be used to illuminate earlier periods for which direct sources are scarce or wholly lacking. It is known, for instance, that some sort of Greek lunar and planetary theory or theories existed at the time of Hipparchus. The appearance of the *Almagest* caused the disappearance from the Greek literature of this body of more primitive doctrine. But some of it found its way to India and has been identified in modern times (see *Neugebauer*, pp. 168 and 178). It is not too much to hope that more pre-Ptolemaic Greek astronomy may turn up in the abundant Muslim literature.

For several of the items in the topical outline of §4, trends can be observed developing during the course of the Islamic period. In the tabular presentation of functions, for instance (topics B through D), there was a steady increase in precision. Thus Khwārizmī and Battānī, at the beginning of the period, tabulate the sine function to three significant places for each degree, whereas Ulugh Beg, at the end of the period, has a sine table with tabular differences, carried to *five significant (sexagesimal) places for each minute of arc.* In the middle of the period there was a tendency to tabulate all the newly-defined functions, the cotangent, the versed sine, and so on. The latest zījes, however, reverted to the practice of tabulating only the sine and tangent.

The computation of ascension tables followed a like pattern. The early zījes have oblique ascension tables for the seven climates only, plus a table for the latitude of the zīj. But, beginning with Ibn Yūnis, a trend established itself to compute a table each for $\varphi = 0°, 1°, 2°, \cdots$, up to some suitable high latitude.

We turn next to a general consideration of planetary equations (topic F). If we restrict ourselves to the Islamic world, the only problem involved is to verify the origin of the Khwārizmī (**21**, §6, F) theory and parameters. Table 2 assembles what are essentially three groups of extreme equations: Ptolemaic, Persian, and Hindu. The very minor differences between those common to Battānī and Zarqālī (from *Delambre*, p. 177) and those of the *Almagest* are typical of the bulk of the zījes—they are unmistakably Ptolemaic. But there is a wide divergence between these and both other groups. At the same time, the parameters of the Shāh Zīj (**30**) as reported by *Ibn Hibintā* (a Christian astrologer of the early Abbasid period), exhibit only trivial differences with those of Khwārizmī, as well as with the "Persian" values as given by *Bīrūnī* (*Risā'il*, III, p. 30 and 54; he includes among these the Shāh Zīj, Abu Ma'shar, (**63**), and Ya'qūb ibn Ṭāriq (**71**)). The incomplete sets of parameters

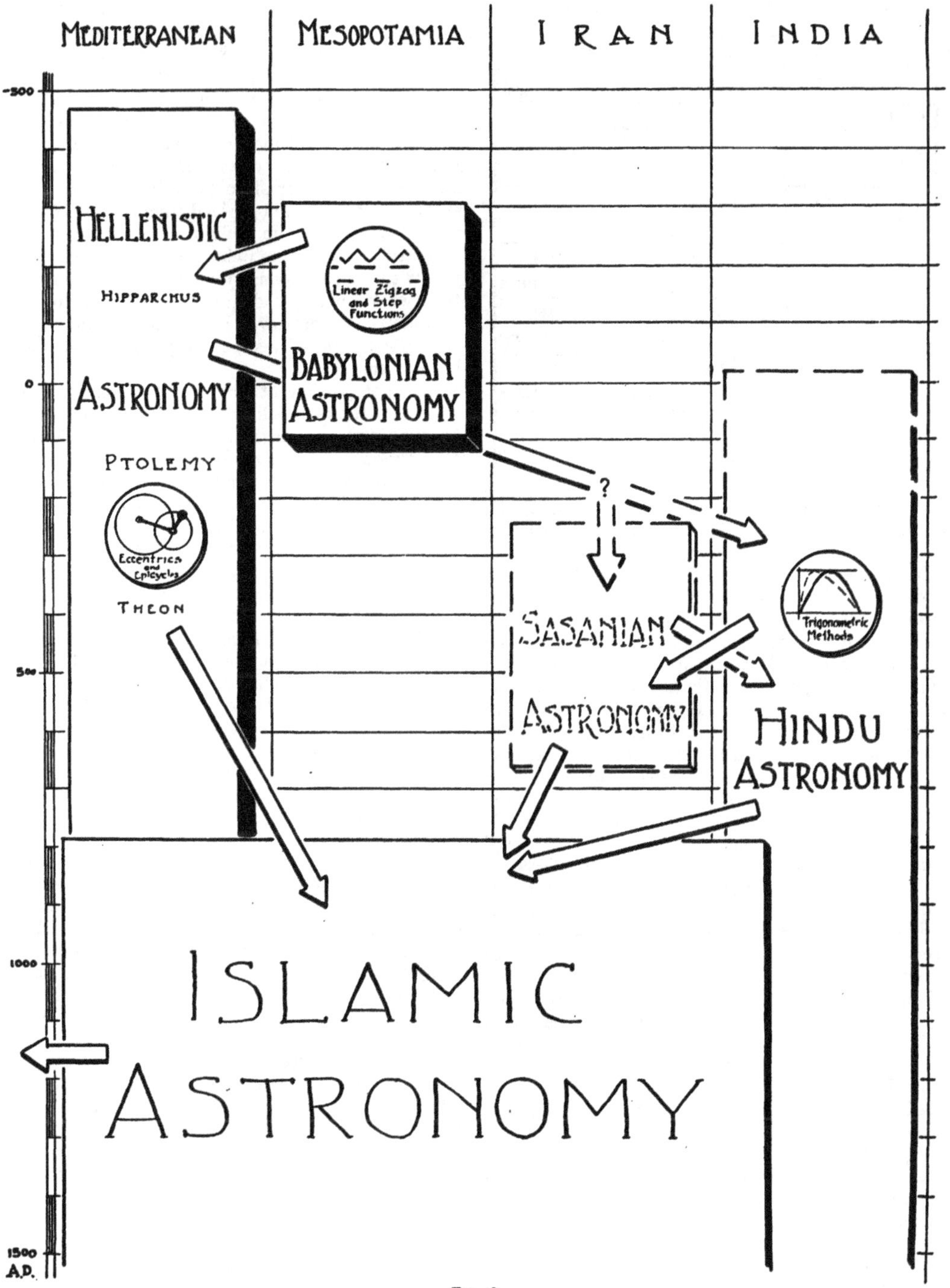
MEDITERRANEAN
MESOPOTAMIA
IRAN
INDIA
-300
0
500
1000
1500 A.D.
HELLENISTIC
HIPPARCHUS
ASTRONOMY
PTOLEMY
Eccentrics and Epicycles
THEON
Linear Zigzag and Step Functions
BABYLONIAN ASTRONOMY
?
SASANIAN ASTRONOMY
Trigonometric Methods
HINDU ASTRONOMY
ISLAMIC ASTRONOMY

FIG. 8

TABLE 2. EXTREME PLANETARY EQUATIONS

	PTOLEMAIC		PERSIAN			HINDU		
	Almagest	Battānī and Zarqālī	Khwārizmī	Shāh	(from Bīrūnī)			Paul the Greek
☉	2;23°	1;59,10°	2;14°	2;14°				
☾	5;1 2;39 13;6	5;1 2;40 13;9	4;56	4;56				
♄	6;31 6;13	6;31 6;13	8;36 5;44	8;36,4 5;43,12	8;37° 5;44	9;28°		 6;22°
♃	5;15 11;3	5;15 11;3	5;6 10;52	5;5,49 10;11,0	5;6 10;52	4;44	5;5°	 11;32
♂	11;25 41;9	11;25 41;9	11;13 40;31	11;11,59 41;9,2	11;12 41;30	11;16	11;25	 40;32
♀	2;24 45;57	1;59 45;59	2;14 47;11	2;12,46 47;11,5	2;13 47;11	2;14	1;16,40	 45;15
☿	3;2 21;2	3;2 22;2	4;2 21;30	4;5,5 22;2,5	4;0 21;30	4;28	4;30	 21;36

from "Hindu zījes" are found in *Bīrūnī, Risā'il*, III, p. 31. A first conclusion is inescapable—Khwārizmī uses pre-Islamic Persian equations. But this leads to a deeper question: did the Iranians invent the theory, or did they take it over from somewhere else, perhaps modifying it in the process. The answer is probably the latter, for although there are differences between the Hindu and the Persian parameters, Khwārizmī's pulsating "apogees" are Hindu, and we have the word of *Bīrūnī* (*Risā'il*, III, p. 54) that Persian astronomical theory came largely from India. As to the origins of the Hindu material, three components can be isolated: the geometric models as pre-Ptolemaic Greek, the "arithmetical methods" as Babylonian, the remainder as indigenous. (*Cf. Neugebauer*, Chap. VI; and Neugebauer, O., "Babylonian Planetary Theory," *Proc. Amer. Philos. Soc.*, vol. 98 (1954), pp. 60–89.)

The situation with regard to planetary latitudes is more complicated. Table 3 shows extreme values of latitudes or latitude components. The first column is from the *Almagest*, the second is a set of values common (with a single exception) to the so-called Mumtaḥan (**51**, §5, G) manuscript and to a set given by *Ibn Hibintā* without attribution. The third column is from Khwārizmī (**21**, §6, G), the pair in the upper line for each planet being maximum and minimum L_1 respectively. The single numbers below are maximum L_2 for each planet. The last column is from the *Sūrya-Siddhānta*.

Again the first column suffices for the bulk of the zījes. They are Ptolemaic. The other three columns differ radically between themselves. With the exception of the moon and Mercury, the entries in column 3 are each two and a half times the entries in column 4. This, combined with a general resemblance between the Khwārizmī tables and the descriptive material in the *Sūrya-Siddhānta* (i, §§68–70), leads to a tentative conclusion that the Khwārizmī latitude theory is Hindu, with or without a Sasanian intermediary.

As for column 2, except for Mercury there is a very close correspondence between its entries for the superior planets and those of the *Almagest*. For Venus the sum of the Ptolemaic components roughly equals the single entry of column 2. Ptolemy's latitude parameters, in contrast to the other planetary constants he used, are not derived from specific observations described by him. They seem to be commonly accepted values current in his time. This, in conjunction with the correspondence between the elements of the two columns, leads to the suspicion that the latitude theory preserved in the Escorial manuscript (**51**) is some sort of vestige of a pre-Ptolemaic Greek latitude theory.

It is not possible at this time to summarize such important topics as eclipse theory or visibility theory for the simple reason that their Islamic ramifications are unknown. We therefore propose to close the section by outlining a program for future study in the field which will include these and other subjects.

The primary need would seem to be for the publication of texts. The Khwārizmī (**21**) and Battānī (**55**) zījes, already published, are from the early part of the period. If the project of an edition of the Khāqānī Zīj (**20**) is carried through, it, plus Sédillot's partial publication of the Sulṭānī Zīj (**12**), would give a sampling of tables from the end of the period. There remains the middle, and by all odds the logical choice would be Bīrūnī's Masudic Canon (**59**). The Arabic text once published, the various parts untouched by Schoy might be attacked in detail and put into some European language, by different individuals, if necessary, and as the occasion offered. A second handbook of the middle period well worth publication is the Sanjarī Zīj (**27**).

In any mention of important texts, the *Bīrūnī Risā'il* should be included. This volume, like all of Bīrūnī's works contains explanations of a great variety of techniques and theories, not only from Islamic sources, but also of Hindu, pre-Islamic, Persian, and Babylonian origins. For instance, *Risā'il*, II, p. 138 has an explanation and numerical example of a method of computing oblique ascensions by an application of what is Babylonian System A (*cf. Neugebauer*, p. 153). This example, named as Babylonian in the text, is the first such instance to be observed in Islamic astronomy. The Oriental Publications Bureau of Osmania University, Hyderabad-Deccan, has performed a notable service for the history of ancient science by making available the text in the original Arabic. This should now be supplemented by a translation into some European language.

Barring an unexpected windfall, it is too much to hope that any of the earliest zījes will turn up in substantially its original form. The Escorial Mumtaḥan

TABLE 3. EXTREME PLANETARY LATITUDES

	1. Almagest	2. Mumtaḥin and Ibn Hibintā	3. Khwārizmī	4. Sūrya-Siddhānta
☾	5;0°	4;30°	4;30°	4;30°
♄	3:2° 3;4°	3;1° 3;6°	1;33° 1;10° 5;0	2
♃	2;4 2;8	2;3 2;9	1;36 0;57 2;30	1
♂	4;21 7;7	4;23* 7;6	2;14 0;27 3;45	1;30
♀	0;10 6;22 2;30	8;56	2;24 0;22 5;0	2
☿	0;45 4;5 2;45	4;18	1;44 0;48 6;15	2

*Ibn Hibintā has 5;23°

manuscript (**51**), with its jumble of unassimilated early and late tables must be regarded as typical. Critical editions of such manuscripts would be an inefficient expenditure of scholarship. Rather it would seem best to use these and all other relevant sources for the preparation of monographs on basic topics presently unexplored. None of the main items on the master outline of §4 have been exhausted, but of them visibility and eclipse theory are largely untouched. Chronology in general is reasonably well covered, but the Chinese-Uighur calendar, on which an extensive literature can be assembled from the later Iranian zījes, offers attractive and unexplored territory. For instance, calendaric computations for this system make standard use of a parabolic interpolation scheme for the solar and lunar equations which has been encountered in no other context.

The surest way of establishing connections between astronomical works is by the comparison of numerical parameters. In the course of the last five years the present writer has picked up a file of roughly a thousand sexagesimal parameters. Of these the mean motion velocities are expressed in various units involving Hijra years, Egyptian years, days, and so on, which conceal the relations between many. Checking, reduction to a standard norm, and publication of the lot would do much to put the classification of zījes on a firm footing.

The foregoing paragraphs have remarked only a few specific projects in a field teeming with unsolved problems. No mention has been made of Islamic material recoverable from Latin, Hebrew, and Byzantine Greek astronomical manuscripts.

The harvest truly is plenteous,
but the laborers are few.

18. CONCLUSIONS

1. In the centuries immediately preceding Islam, at least one astronomical handbook, and probably more, were computed and used in Sasanian Iran. Some observational activity is attested (**14**, Chap. 8, p. 124, Leiden copy). The available evidence suggests that the Sasanian astronomy was under strong Hindu influence.

2. In the eight centuries beginning with A.D. 700 well over a hundred different zījes were produced. Of these, more than twenty were based at least partially on observations made by their respective authors. Others differed from prototypes only to the extent, say, that mean motion tables were recomputed for a different epoch and calendar.

3. In the great majority of these tables the basic theory is that of Ptolemy, although with some improvement of parameters. The main Islamic contributions were in trigonometrical, computational, and observational technique.

4. A minority of zījes were based on Hindu or pre-Islamic Iranian theory. Of these, the only one preserved is the Khwārizmī Zīj which has

(*a*) Planetary equations from the Sasanian Shāh Zīj (**30**),

(*b*) Planetary latitudes not presently identifiable, but non-Ptolemaic and probably Hindu,

(*c*) Planetary stations from the *Almagest*, hence irreconcilable with (*a*) and (*b*) above,

(*d*) Two values of the length of the sidereal year, one a common Hindu parameter used in the Brahmasiddhānta, the Siddhānta Śiromaṇi and the Arabic Sindhind (**28**), the other a Persian value.

BIBLIOGRAPHY AND ABBREVIATIONS

Almagest. Syntaxis mathematica, ed. J. L. Heiberg, 2 v., Leipzig, 1898–1903. German translation by K. Manitius, 2 v., Leipzig, 1912–1913.

Bankipore. Catalogue of the Arabic and Persian Mss. in the Oriental Public Library at Bankipore, vols. XI and XXII, Calcutta, 1927 and 1937.

Berlin. Ahlwardt, W., Verzeichniss der arab. Hss. der königlichen Bibl. zu Berlin, vol. 5, Berlin, 1893.

Bīrūnī, Chron. The Chronology of Ancient Nations . . . , transl. and ed. by E. C. Sachau, London, 1897.

Bīrūnī, India. Alberuni's India, edited and translated by E. C. Sachau, 3 v., London, 1887 and 1910.

Bīrūnī, Risā'il. Risā'il al-Bīrūnī, Oriental Publ. Bureau, Osmania Univ., Lallaguda, Hyderabad-Deccan, 1948.

Bīrūnī, Tafhīm. Wright, R. R., Elements of Astrology by al-Bīrūnī, London, 1934.

Bodl. II, 1. Uri, Bibliothecae Bodleianae codicum manuscriptorum orientalium, . . . catalogus, . . . Pars Prima, Oxonii, 1787.

Bodl. II, 2. Pusey, Bibliothecae Bodleianae codicum manuscriptorum orientalium catalogi partis secundae volumen secundum arabicos complectens, Oxonii, 1835.

Bodl. Pers. (Sachau and) Ethé, Catalogue of the Persian . . . Mss. in the Bodleian Library, Part I, The Persian Mss., Oxford, 1889.

Bouché-Leclercq. L'Astrologie Grecque, Paris, 1899.

Brit. Mus., II, 2. Catalogus Codicum Orientalium Musei Britannici, Pars Secunda, Codices Arabicos Amplectens, Londini, MDCCCLII.

Brit. Mus., Suppl. Rieu, C., Supplement to the Catal. of the Arabic Mss. in the British Museum, London, 1894.

Browne. A descriptive Catalogue of the Oriental Mss. belonging to the late E. G. Browne, by Edward G. Browne. Completed and edited with a Memoir of the Author and a Bibliography of his writings by Reynold A. Nicholson, Cambridge, Univ. Press, 1932.

Būhār. Catalogue raisonné of the Būhār Library, Calcutta, 1923.

Cairo. Al-juz' al-khāmis min fihrist al-kutub al-'arabiya al-maḥfūẓa b'il-kutubkhāna al-khadiwiya al-miṣriya, Cairo, 1308 (A.H.).

Cambr. Browne, E. G., A hand-list of the Muḥ. Mss. . . . in the Library of the Univ. of Cambridge, Cambridge, 1900.

Caussin. Caussin de Perceval, Le Livre de la grande Table Hakémite, . . . par Ebn Iounis . . . , Notices et extraits des mss. de la bibl. nationale . . . , tome septième, Paris, an XII de la république.

CCAG. Catalogus Codicum Astrologorum Graecorum, 12 v., Brussells, 1899–1953.

Curtze. Urkunden zur Gesch. der Trig. . . . , Bibl. Math., 1 (1900), pp. 321–416.

Delambre. Hist. de l'astron. du moyen age, Paris, 1819.

Escorial, I. Bibl. Arabico-Hispana Escurialensis opera M. Casiri, 2 v., Matriti, 1760–1770.

Escorial, II. Derenbourg, Les mss. arabes de l'Escurial, Tome II, fasisc. 2, Paris, 1941.

Fihrist. Al-Fihrist li-Ibn an-Nadīm, Cairo, 1348 (A.H.).

Fihrist, transl. Suter, H., Das Mathematiker-Verzeichniss im Fihrist des Ibn Abū Ja'kub an-Nadīm . . . , Abhand. zur Gesch. der Math., Sechstes Heft, Leipzig, 1892.

GAL. Brockelmann, C., Gesch. der arabischen Litteratur, 2 v. (2d ed.) and 3 suppl. v., Leiden, 1943.

Ginzel. Handbuch der mathematischen und technischen Chronologie . . . , Leipzig, 1906.

Handy Tables, or *Tables faciles.* Halma, Commentaire de Théon d'Alexandrie sur le livre III de l'Almageste de Ptolemée, 3 parts, Paris, 1822–1825.

Hāshimī. 'Ali bin Suleimān, Kitāb 'ilal az-zījāt, *Bodl. II, 1*, Ms. 879, 4 (Seld. A. 11).

d'Herbelot. Bibliotheque Orientale . . . , Maestricht, 1776.

HKh. Flügel, Gustav, Kashf aẓ-Ẓunūn . . . , Lexicon bibliographicum et encyclopaedicum a Mustafa ben Abdallah . . . Haji Khalifa celebrato compositum, 7 v., Leipzig and London, 1835–1858.

Ibn Hibintā. Ms. (Munich) Cod. arab. 852, al-mughni fi'n-nujūm.

Ibn al-Qifṭī. Ta'rīḫ al-Ḥukamā', . . . herausgegeben von Dr. Julius Lippert, Leipzig, 1903.

Ind. Off. Loth, O., Cat. of the Arabic Mss. in the Libr. of the India Office, London, 1877.

Ind. Off. Pers. Ethé, H., Cat. of Pers. Mss. in the Library of the India Office, Vol. I, Oxford, 1903.

Kary-Niyazov. T. N., Astronomicheskaya Shkola Ulugbeka, Moscow and Leningrad, 1950.

Khwar. Björnbo and Suter, Die astronomischen Tafeln des . . . al-Khwārizmī . . . , Copenhagen, 1914.

Kennedy 1. Kennedy, E. S., Parallax Theory in Islamic Astronomy, Isis, Vol. XLVII (1956), pp. 33–53.

Kennedy 2. Kennedy, E. S., An Islamic Computer for Planetary Latitudes, Jour. Amer. Or. Soc., vol. 71 (1951), pp. 13–21.

Knobel, Chron. The Chronology of Star Catalogues, Memoirs of the Royal Astr. Soc., vol. XLIII (1875–1877), pp. 1–23.

Knobel. Ulugh Beg's Catalogue of Stars, Washington, 1917.

Krause. Krause, M., Stambuler Handschriften islamischer Mathematiker, Quellen und Studien zur Gesch. der Math. . . . , Abt. B, Bd. 3, Berlin, 1936, pp. 437–532.

Lee. Lee, Samuel, Notice of the Astronomical Tables of Mohammed Abibekr al Farsi, . . . , Trans. of the Cambridge Philosophical Society, vol. I, 1822, pp. 249–265.

Leiden. Dozy, de Jong, de Goeje et Houtsma, Catalogus codicus orientalium bibliothecae academiae Lugduno Batavae, VI v., Lugd. Bat. 1851–1877.

Meshed. Catalogue, Oktā'ī, Fihrist-i kutub-i kitābkhāneh-i mubārakeh-i ostān-i quds-i riḍavī, 3 v., 1345 (A.H.).

Millás Vallicrosa. Millás Vallicrosa, J., El libro de los fundamentos de las Tablas astronomicas de R. Abraham ibn 'Ezra, edición crítica, con introducción y notas, Madrid-Barcelona, 1947.

Millás Vallicrosa 2. Millás Vallicrosa, J., Estudios sobre Azarquiel, Madrid-Granada, 1943–1950.

Munich. Aumer, J., Die arab. Hss. der k. Hof- und Staatsbibliotek in Muenchen, München, 1866.

Nallino. Nallino, C. A., Raccolta di Scritti editi e inediti, vol. V, Astrologia, Astronomia, Geografia, Rome, 1944.

Nallino, Batt. Al-Battānī sive Albatenii Opus Astronomicum, 3 v., Milan, 1899–1907.

Neugebauer. Neugebauer, O., The exact sciences in antiquity, Copenhagen, 1951.

Paris. Catalogue des Mss. arabes par M. le Baron de Slane, Paris, 1883–1895.

Princeton, Pers. Descriptive Catalog of the Garrett Collection of Persian, Turkish, and Indic Manuscripts in the Princeton University Library, Princeton, 1939.

Rāmpūr. Fihrist Kitāb 'Arabī, Catalogue of Arabic Books in the Rāmpūr State Library, 1902.

Schoy, Geogr. Schoy, C., Aus der astronomischen Geographie der Araber, Isis, vol. V, (1923), pp. 51–74.

Schoy, Mas'ūdī. Die trigonometrischen Lehren des persischen Astronomen Abū'l-Raiḥān Muḥ. ibn Aḥmad al-Bīrūnī, dargestellt nach al-Qānūn al-Mas'ūdī, Hannover, 1927.

Schoy, Trig. Beiträge zur arabischen Trigonometrie, Isis, vol. V, (1923), pp. 364–399.

Sédillot. Sédillot, L. A., Prolégomènes des tables astronomiques d'Oloug-Beg . . . , Paris, 1847.

Seemann. Seemann, H. J., Die Instrumente der Sternwarte zu Marāgha . . . , Sitzungsberichte der phys.-med. Sozietät zu Erlangen, Bd. 60 (1928), pp. 15–126.

Steinschneider, Études. Études sur Zarqālī . . . , (Continuazione), Bull. di Bibl. e di Storia delle Scienze Math. e Fisiche, pubbl. da B. Boncompagni, Tomo XX, Roma, 1887, pp. 1–36.

Sūrya-Siddhānta. Burgess, E., Translation of the Sūrya-Siddhānta, a textbook of Hindu astronomy . . . , reprinted from the edition of 1860, Univ. of Calcutta, 1935.

Suter. Suter, H., Die Mathematiker und Astronomen der Araber und ihre Werke, Abhand. zur Gesch. der Math. Wiss. 1, X Heft, Leipzig, 1900.

Suter, Nachtr. Nachträge und Berichtigungen zu "Die Math. und Astr. . . . ," Abhand. zur Gesch. der Math. Wiss., XIV Heft, Leipzig, 1902.

Taqīzādeh. Taqīzādeh, S. H., Gāh shumārī dar Īrān-i qadīm, Tehran, 1316 (Hijrī-i shamsī).

Tehran. Etessami, Y., Catalogue des mss. pers. et ar. de la bibl. du Madjless, 2 v., Tehran, 1933.

Tetrab. Ptolemy, Tetrabiblos, ed. . . . by F. E. Robbins, London and Cambridge (Mass.), 1940.

Vatican. Bibliothecae Apostolicae Vaticanae codd. ms. catalogus, Romae, 1766.

Vat. V. Levi della Vida, Elenco dei Manoscritti arabi islamici della Biblioteca Vaticana . . . , Città del Vaticano, 1935 (Studi e Testi 67).

Wüstenfeld. Wüstenfeld, F., Die Übersetzungen arabischer Werke in das Lateinische . . . , Göttingen, 1877.

ZDMG. Zeitschrift der Deutschen Morgenländischen Gesellschaft.

INDEX

All numbers below are zij serial numbers. Where a number is printed in boldface type the name opposite it is either that of the author or that of the zij itself. Initial *al-*, *abu*, and *ibn* have been left off of names.

'Abd al-Karim, **23**
'Abd ar-Raḥmān, **107**
Abharī, **40**, 56
Adamī, **8**, **18**
'Aḍudī, **70**
Aḥmad ibn Mūsā, **92**
'Alā'ī, 29, 40, **42**, 84
'Alā'ī ar-Raṣadī, **84**
A'lam, 9, 23, **70**
Albategnius, **55**
Albumasar, **63**
Almagest, **X207**
Amad, **5**
Amājūr, **8**, **67**, **78**, **79**, **90**
Ammonius, **X213**
Arabic zīj, 15
Arkand, 16, **X214**
Āryakhaṇḍa, **X214**
A'sharī, **X210**
Ashrafī, **4**
'Āṣim, 100
Asṭurlābī, **52**
Athīr, **40**, **56**, **83**
Athīrī, **56**, 81
Azarchiel, **24**

Badī', **8**
Baghdādī, **3**
Bākū'ī, **23**
Bāligh, **7**, 9
Balkhī, **63**
Baṭlamyūs, **X207**
Battānī, 9, 24, 44, 49, **55**, 65
Bāzyār, **68**
Bīrūnī, 2, 9, 15, 21, 31, 45, **59**, 63, 73, 77, 82, 100, 101, 103, 104, 105, X214
Brahmagupta, **X214**
Bukhārī, **25**, 32, 35
Būzjānī, 29, **73**

Cassini, **X208**
Cyriacus, **81**

Dahhān, **89**
Damascus Zīj, 15
Dānishī, **49**
Dīnawārī, **19**
Durr al-Muntakhib, **81**

Eumathius, **X213**

Fahhād, 23, **53**, 54, **58**, **62**, **64**, **84**, 99
Fākhir, **44**
Fāriqī, **47**
Fārisī, 23, 44, **54**, 65, 84, 93, 94, 95, 97, 98, 99
Farrukhān, **104**
Fazārī, 2, **45**

Gharnāṭī, **109**
Gurgānī, **12**

Ḥabash, 8, 9, **15**, **16**, 23, **39**, **68**
Hā'im, **48**, **66**
Ḥākimī, **14**

Ḥalabī, **34**
Hamdānī, **69**
Handy Tables, **X205**
Ḥārith, **61**
Harqan, **X206**
Hārūn bin 'Alī bin Yaḥyā, **102**
Hārūnī, **101**
Hāshimī, 63, **82**
Ḥawālfa'sī (?), 55
Hazārāt, **63**
Hindisi, **X211**
Hurmuzī, **33**
Ḥusām-i Sālār, **32**

Ibrāhīm, **2**
Ikhtiyārī, **83**
Īlkhānī, **6**, 10, 20, 35
'Ilm al-Falak, **37**
Intikhābī, **X220**
Isḥāq ibn Ḥunein, **94**

Jadīd, **11**, **12**, **13**
Ja'far, **106**
Jahen, **76**
Jai Singh, **X203**
Jamāl ad-Dīn, **3**
Jāmi', 9, **10**, **X220**
Jamshīd, 10, 20, **88**
Jawharī, 23, **99**
Juhanī, **76**

Kabīr, **14**, **18**, **46**
Kāfī, **103**
Kamad, **5**, **66**, **72**
Kāmil, **48**, **49**, **82**
Kandakātik, **X214**, **X218**
Kankah, 71
Karaṇasāra, **X219**
Karaṇatilaka, **X217**
Kāshī, 20, **88**
Kassawṭuh, **X215**
Kaur, **72**
Khāliṣ, **78**
Khaṇḍakhādyaka, **X214**
Khāqānī, 20, **80**
Khayyām, **22**
Khāzin, **X200**
al-Khāzinī, **27**
Khujandī, **61**
Khwārizmī, **21**
Kurkānī, 12
Kūshyār, **7**, 9, **44**

Lubūdi, **86**, **87**

Madā'inī, 18
Maghribī, **41**, **108**
Māhānī, **98**
Maḥfūẓ, **3**
Maḥlūl, **71**
Maḥmūdī, **52**
Majdī, **36**
Majisṭī, **73**, **77**
Majrīṭī, **21**
Mājūr, **8**

Makka, **X202**
Malikshāhī, **22**
Mamarrāt, **67**
Ma'mūn, **50**
Ma'mūnī, 15, **51**
Marvarūdī, 16, 23, **85**, **97**
Ma'shar, 9, **63**, 106
Masīḥ, **109**
Maslama, **21**
Mas'ūdī, **59**
Masudic Canon, **59**
Miṣbāḥ, **31**
Mu'addal, **62**
Mufannun, **74**
Mufīd, **X209**
Mufrad, **65**
Mughnī, **64**
Muḥammad Shāhī, **X203**
Muḥaqqaq, **35**
Muḥī ad-Dīn, **108**
Muḥkam, **53**
Mujarrab, **51**
Mukhtār, **57**
Mukhtari', **31**
Muktaṣar, **17**, **85**, **105**
Mulakhkhaṣ, **40**
Mumtaḥan, 15, **51**, **54**, 85, 87
Mumtaḥan al-'Arabī, **54**
Mumtaḥan al-Khazā'inī, **54**
Mumtaḥan al-Muẓaffarī, **54**
Muqarrab, **87**
Muqtabas, **66**
Mūsā bin Shākir, **91**
Mushtamil, **1**
Muṣṭalaḥ, **47**
Mustaufī, **58**
Mustawī, **58**
Mu'tabar, **27**
Mu'tadil, **62**
Muthannā, **X212**
Muzannar, **79**

Nairain, **X216**
Nairīzī, **46**, 63, **75**
Nasawī, **44**
Naṣīr ad-Dīn aṭ-Ṭūsī, **6**, 20, 32, 35, 108
Naṣr Manṣūr, **77**
Naṣrānī, **X215**
Naẓm al-'Iqd, **18**
Nihāwandī, **1**
Niẓām al-A'raj, **42**

Probata, **51**
Ptolemy, **X207**

Qānūn, 15, 59, **X205**, **X213**
Qasīnī, **X208**
Qirānāt, **63**
Qiriāqus, **81**
Quṭb ad-Dīn, **13**, **25**

Raṣadī, **84**
Reiḥān, **59**
Riḍwānī, **13**